The Lost Frontier

The Lost Frontier

Water Diversion in the Growth and Destruction of Owens Valley Agriculture

Robert A. Sauder

The University of Arizona Press . . Tucson & London

The University of Arizona Press

∞ This book is printed on acid-free, archival-quality paper.
Manufactured in the United States of America

99 98 97 96 95 94 6 5 4 3 2 1

Library of Congress Cataloging-in-Publication Data

Sauder, Robert A.
The lost frontier : water diversion in the growth and destruction of Owens Valley agriculture / Robert A. Sauder.
p. cm.
Includes bibliography and index.
ISBN 0-8165-1381-3 (alk. paper)
1. Land settlement — California — Owens River Valley — History.
2. Agricultural colonies — California — Owens River Valley — History.
3. Water transfer — California — Owens River Valley — History.
4. Owens River (Calif.) — Water rights — History. 5. Water-supply — California — Los Angeles — History. I. Title.
HD1516.U7C27 1994
338.1'09794'87 — dc20 93-41380
CIP

British Cataloguing-in-Publication Data
A catalogue record for this book is available from the British Library.

To Rose

Contents

Figures

Tables

Preface

As a resident of southern California for the first twenty-four years of my life, I have long held an interest in the Owens Valley–Los Angeles water conflict. Yet, after having read most of the voluminous literature that is the product of many years of controversy, I concluded that I still knew very little about the Owens Valley's past: where the early pioneers came from, what hardships they encountered in this arid environment, what kind of farming system they adopted, and what adjustments they needed to make to survive in such an isolated region of California. In short, I wanted to learn more about the processes and patterns of environmental adaptation and settlement in the valley. Only then would I have a better sense of what had been lost as a result of Los Angeles's intrusion.

Several years ago I began examining the details of the Owens Valley's overlooked past. My primary objective was to reconstruct the patterns of rural livelihood and settlement that once were found there and that have largely been expunged from the valley floor as a result of Los Angeles's activities. Even though the focus of my research was this little-known (prior to the water issue) region of California, once I was into the project it became apparent that the Owens Valley was a microcosm of broader processes and patterns that characterized much of the intermountain region of the arid West. Where relevant, therefore, I have made every effort to link my findings to a larger framework concerning Western settlement generally. In so doing, I also hope to illuminate the relationship between the Owens Valley's past and the future of existing rural communities scattered throughout the arid West.

Chapters 1 and 2 set the stage for the chapters that follow. Even though much of the physical and cultural information contained in Chapter 2 is already well known, it provides the needed background for the subsequent chapters. It is the analysis based on material from heretofore untouched sources regarding the settlement of the Owens Valley that I hope will be viewed as the primary contribution of this study. In this connection I consulted several sets of data in order to piece together the evolving patterns of livelihood and settlement in the valley. I examined field notes and plats of the original federal land survey of the Owens Valley to determine the degree to which they helped guide the initial course of settlement in the region. Tract book data from the Independence District Land Office reveal how settlers and land speculators used the various federal land laws to take up public domain in the valley. Master title township plat maps show the land which has been patented to private ownership; their companion historical indexes provide a chronological record of patented and canceled entries in each township. Federal manuscript census schedules and the U.S. census of agriculture for Inyo County reflect changing patterns of livelihood in the valley. The region's two early weekly newspapers, the *Inyo Independent* and the *Inyo Register,* proved to contain a wealth of contemporary information when used with caution to separate fact from fiction (that is, boosterism). Finally, an examination of historical maps and records housed in the Engineering Division of the Los Angeles Department of Water and Power and numerous field excursions into the Owens Valley in an effort to view the land through the eyes of the pioneers also proved invaluable.

Equally important for the completion of this project were the many people who contributed their time to help ensure that the story would be told. The individuals who assisted me are too numerous to include every one here, and I do not in fact know all their names. To those who remain unmentioned, I extend my heartfelt thanks. But a few especially gracious persons must be acknowledged, for without their assistance I would still be tracking down potential leads. To my wife, Rose, also a geographer, I extend my warmest appreciation. She has been an untiring companion, contributing to nearly every phase of the project from research in volumes of old newspaper, manuscript census, and tract book microfilms to the many cross-country forays into the Owens Valley and to the final reading of the manuscript. Two especially knowledgeable individuals based in the Owens Valley—Dave Babb, Range and Wildlife Management Specialist for the Los Angeles Department of Water and Power, Bishop Office, and Bill Michael, Administrator of the Eastern California Museum in Independence – shared with me many insights and source materials that I would never have uncovered on my own. I shall not forget how they would consistently drop whatever they were doing to lend a hand whenever I stopped by unannounced. During the early phases of my research, and prior to her retirement, Norma Aguirre of the Water Engineering Design Division, Los Angeles Department of Water and Power, graciously helped me locate and copy land use, land ownership, and other relevant historical maps housed in her division. Tami Mistrick and Barbara Trapido did an outstanding job of preparing my original maps for publication.

I also wish to extend my gratitude to the staffs of the Huntington Library, the Inyo Free Library, the California Room of the California State Library, the Public Information Room of the Bureau of Land Management's California State Office in Sacramento, and the U.S. Forest Service and Soil Conservation Service in Bishop. Finally, on numerous occasions the University of New Orleans Research Council and the College of Liberal Arts provided generous support for this project, including several summer research stipends, thereby extending both free time and financial backing to pursue my research. It is with sincere gratitude that I acknowledge the university's confidence, and patience, in my ability to carry this project to completion. I wish to emphasize, however, that while the persons and institutions recognized above graciously facilitated my research the interpretation presented is mine alone. Any errors that may have crept into this work are therefore solely the responsibility of the author.

The Lost Frontier

1. Colonizing An Arid Frontier

> Only the desperate farmer, shut out of opportunity in more favored places, would consider the desert for a dwelling. . . . But there were other Americans who came ready to believe in miracles, who began to insist it was precisely that deficiency [in water] which made the West so promising.
>
> —Donald Worster, *Rivers of Empire*

Agricultural colonization, according to Donald Meinig, is a process that imposes radical change on any region.[1] The colonization of the Great Basin frontier of the western United States was no exception.[2] The process of colonizing the arid intermountain West, however, is only vaguely understood. Recent works by Donald Worster and Marc Reisner tell the story of water resource development and Western settlement in broad terms, but little *detailed* work chronicling pioneer settlement in this dry region of the country has been undertaken.[3] Numerous studies focus on the Mormon colonization of the eastern Great Basin, but in contrast to most of the pioneering efforts in the intermountain West, Mormon settlement was a communal endeavor heavily influenced by the church.[4] It therefore lacked the spontaneity of individual effort more commonly associated with the pioneering process. Marshall Bowen's series of articles detailing pioneer settlement in northeastern Nevada provides insights into the processes, patterns, and problems associated with dryland farming in the Great Basin.[5] But dryland farming was generally an ephemeral activity in the region.[6] More permanent Great Basin settlements were associated with the establishment of early mining camps and the irrigation agriculture that developed in response to mining activities.[7] This study focuses on California's Owens Valley, which was part of the latter category of settlement in the Great Basin frontier.

Since this frontier, limited by aridity and fragmented by topography, was not a simple extension of the continually westward moving frontier of the humid prairies and subhumid plains, it is assumed that its settlement processes did not mirror those of the humid East.[8] Instead, because of the region's vast distances, rugged topography, dry environments, and Pacific Slope influences, it is generally believed that this was a distinct frontier with only ephemeral links to the East. Certainly, people had to overcome numerous environmental constraints if the agricultural colonization of the arid frontier were to succeed, but the majority of pioneers who migrated to the region came from the East, often via the Western mines. To what extent did they attempt to transplant in the arid West patterns of human organization common to the humid East, and what adaptations and adjustments were necessary for settlement to succeed?

The onset of colonization in the Owens Valley in the early 1860s is one of the earliest attempts by *individual* pioneers to settle the western Great Basin. Historically, the Owens Valley's notoriety derives not from early settlement but from the capture, early in this century, of the valley's surplus water supply by the rapidly growing city of Los Angeles. When water from the Owens River first spilled into the Los Angeles Aqueduct in 1913, the lifeblood of the Owens Valley began to drain away, though it would be another decade before the full impact of water withdrawal would be felt in the valley. Much has been written about the manner in which Los Angeles acquired Owens Valley water and subsequently diverted it south to quench the thirst of the metropolis, and some of the distortions or embellishments of the facts provide exciting reading.[9] But overlooked in the voluminous literature on the controversy is a thorough examination of the patterns of rural livelihood and land use that had evolved in the valley prior to the Los Angeles water diversion.[10] The purpose of this analysis of colonization efforts in the Owens Valley is to illuminate the processes, patterns, hardships, and adjustments associated with colonizing this arid frontier, for it is likely that similar processes operated in other Great Basin communities. By reconstructing the patterns of human organization in the Owens Valley in the context of colonization, this study also attempts to uncover the nature and extent of agricultural developments in the region at the time that its water was siphoned off to the thirsty metropolis to the south.

Reconstructing the relationships between Owens Valley settlers and their environment is not easy, for the Owens River diversion subsequently led to farmland abandonment and the disappearance of most traces of rural settlement formerly found in the valley. After experiencing more than sixty years of colonization efforts, the Owens Valley in the mid-1920s became a virtual colony of Los Angeles. As farmers left the valley, abandoned farmhouses were bulldozed by Los Angeles, and once-productive fruit orchards and grain and alfalfa fields were invaded by desert scrub. A few relict landscape features reflecting past rural activities, however, linger in the valley. Exotic vegetation planted by former settlers, such as lombardy poplar and black locust trees, help ward off the valley's heat and aridity; unlined irrigation ditches, some still carrying water, spread across the valley floor; abandoned concrete silos stand forlorn against the mountain backdrop; and a widely scattered checkerboard of emerald-green fields supporting alfalfa, a mere remnant of a much larger acreage, prevails to this day.

Two additional purposes are also woven into this narrative. One is to provide a yardstick against which recent heated debates in the Owens Valley, particularly those dealing with environmental questions, may be measured. The dynamics of landscape alteration resulting from past human actions remain largely unexplored while the debate over the more recent impact of Los Angeles's activities in the valley wears on. Yet the effects of the city's actions cannot be adequately assessed without some knowledge of the cumulative impact of past events. The second purpose is to examine the Owens Valley case in the larger context of the relationship between agriculture and urban growth in the West as a whole. Today, the rivers of the West are generally fully appropriated, and new and expanding municipal and industrial uses in the region can be accommodated only through transfers from established agricultural uses. As the Owens Valley experience demonstrates so dramatically, the shift of water away from farming has significant implications for those who are tied to agriculture. The Owens Valley's fate, in which agricultural water was siphoned off to serve municipal needs without adequately safeguarding the rights of valley citizens, provides important insights into options for preventing the piecemeal disassembly of Western agriculture by growing municipalities. The Owens Valley, then, not only reveals much about the arid West's past, it also allows us to foretell, and perhaps to guide, the region's destiny.

2. In the Shadow of the Sierra

> Before us, and far beneath us lay Owens Valley, the desert Inyo Mountains beyond, dry and forbidding. . . . The Sierra Nevada catches all the rains and clouds from the west – to the east are deserts – so, of course, this valley sees but little rain, but where streams come down from the Sierra they spread out and great meadows of green grass occur. . . . Yet these meadows comprise not over one-tenth of the valley – the rest is desert.
>
> – William H. Brewer, *Up and Down California* (1864)

> To understand the fashion of any life, one must know the land it is lived in and the procession of the year.
>
> – Mary Austin, *The Land of Little Rain* (1903)

The Natural Setting

The Owens Valley, located in east central California on the western margin of the Great Basin, is situated in the rain shadow between the lofty, glaciated peaks of the Sierra Nevada and the barren, forbidding White-Inyo Mountains (Figs. 2.1 and 2.2). It is a region of outstanding natural beauty, with topographic relief matched by few other places in the western United States. The 8,000-foot escarpment of the Sierra, whose crest averages 12,500 feet and culminates in Mt. Whitney at 14,495 feet, dramatically defines the western wall of the valley. To the east, the crest of the White-Inyo Mountains attains an average elevation of 10,000 feet. The west face of the White-Inyo range is striking in its nearly linear trend, while on the opposite side of the valley the Sierra front is offset about eight miles to the west, forming Round Valley, a branch of the main valley, northwest of Bishop (Fig. 2.1).

The Sierra Nevada took shape about 210 million years ago when an ancient sea covered the western portion of the continent. At that time, the Sierra batholith, a mass of molten material of unknown depth, began to cool and crystallize into granite, a process that took more than

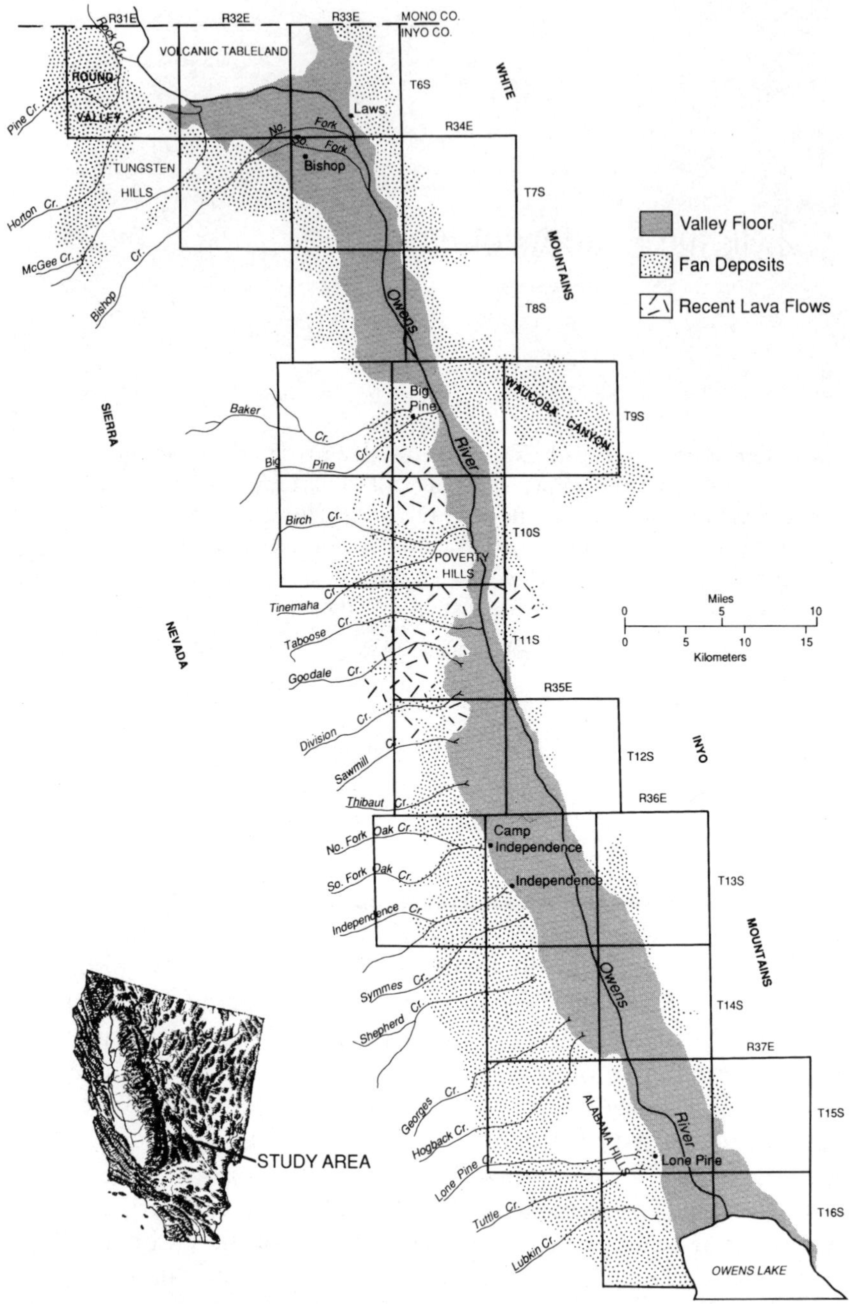

Figure 2.1. The Owens Valley in regional context. The township grid identifies the study area on this and succeeding maps. (Reproduced from Sauder, "Patenting An Arid Frontier," 546, by permission of the Association of American Geographers)

Figure 2.2. An aerial view of the Owens Valley looking south. The settled area in the foreground is Bishop. The Sierra Nevada borders the valley on the right, the White Mountains on the left. The volcanic tableland is at the bottom of the photo.

100 million years to complete.[1] About 130 million years ago, an arching of the earth's crust occurred along the axis of this batholith, causing a low mountain range to emerge above the surface of the sea. The Sierra Nevada entered a phase of deep erosion, exposing its granitic core and wearing the range down to low hills. In a series of subsequent uplifts that began 20 million years ago, the Sierran block was tilted to the west as it rose, and the eastern flank broke into a series of eastward-tilted basin and range blocks. The most significant period of uplift began about 3 million years ago, when the range was raised to its present height, and the wedge-shaped trough that now forms the Owens Valley fell to as much as 5,000 feet below sea level.[2] The crustal activity that formed the Owens Valley has continued to recent times. During the Owens Valley Earthquake of March 1872 – reputed to be the most severe California earthquake in recorded history – the eastern mountain

Figure 2.3. A view of the alluvial slope at the base of the Sierra Nevada north of Independence.

front rose thirteen feet in just a few minutes and moved to the south as much as twenty feet.[3]

As the mountain blocks rose and the valley floor subsided, streams and other agents eroded the escarpments that formed the valley walls and transported their debris into the valley. Today the valley fill is estimated to be 8,000 feet thick in places.[4] The great alluvial fans along the base of the Sierra Nevada average one to seven miles in width, and south of Poverty Hills they merge to form a continuous apron (Fig. 2.3). The younger fans at the base of the White-Inyo Mountains are small by comparison, but pronounced. Those along the foot of the White Mountains also have coalesced to form a continuous alluvial slope.

While earth movements were occurring along fault zones in the region, volcanic vents and fissures were pouring out large amounts of lava and pyroclastic debris, much of which now lies buried beneath the valley floor. About a million years ago, a series of ash flows filled the north end of the Owens Valley which, when consolidated, formed an

Figure 2.4. An aerial view of Round Valley (foreground) and Bishop (background).

800-foot-thick rhyolitic volcanic tableland (Figs. 2.1 and 2.2). More recent eruptions of basaltic materials also occurred intermittently along the margins of the Owens Valley south of Big Pine, forming a conspicuous volcanic field of basalt surrounding Red Mountain and Crater Mountain, both of which are cinder cones. With the exception of these recent volcanic deposits, the flatness of the valley floor is interrupted by only three groups of hills – the Tungsten, Poverty, and Alabama hills – each projecting as a knob of bedrock above the valley fill (Fig. 2.1).

At the foot of the alluvial slopes, the floor of the Owens Valley averages between four and fifteen miles across. It reaches its greatest width and highest elevation in the north, in the Round Valley–Bishop region (Fig. 2.4). At the base of the volcanic tableland, the valley floor descends from an elevation of 4,800 feet in Round Valley to 3,600 feet on the dry bed of Owens Lake, 75 miles to the south. Glacial Lake Owens is part of a chain of interconnected Pleistocene lakes that formerly occupied parts of Death Valley and the Owens, Searles, and Panamint basins (Fig. 2.5).[5] Glaciers, which scoured the slopes of the Sierra Nevada during the Pleistocene, poured an enormous volume of water into the Owens Valley when they subsequently melted and retreated upslope. As the level of Owens Lake rose, it overflowed to the south, spilling water into the integrated system of drainage basins. Then as the lake level gradually receded about 4,000 years ago, Owens Lake was separated from the rest of the lake system, becoming a closed basin.[6] Old beach deposits indicate that the lake at its highest level was as much as 250 feet deep and extended at least ten miles north of Lone Pine, covering an area of 220 square miles.[7]

The Owens River, which rises in the Sierra Nevada in Mono County to the north, drains the Owens Valley depression. Emerging from a gorge that it has cut into the volcanic tableland, the river flows eastward along the southern edge of lava, following the general tilt of the land (Figs. 2.2 and 2.4). Northwest of Laws, where the valley begins to slope gradually southward, the Owens River turns south and generally meanders along the east side of the valley (Fig. 2.6). Originally it emptied into Owens Lake, before Los Angeles diverted the river water. The streams draining the White-Inyo range are intermittent and convey little runoff. On the west, numerous perennial Sierra Nevada creeks, fed by snow and glaciers, flow out of their canyons, across the alluvial apron they have deposited, and on toward the Owens River. Many of these streams, particularly those south of Poverty Hills, sink into the loose detrital material before reaching the Owens River. They therefore contribute to the vast storage reservoir housed in the porous valley fill beneath the surface.[8] The granite cores of the bordering mountains and the bedrock south of Owens Lake serve as impervious barriers against the escape of these underground waters, making the Owens Valley a watertight basin.[9] But while the snow-fed streams of the Sierra are responsible for the valley's abundant water resources, the Sierra Nevada also casts a rain shadow across the region as it captures and depletes moisture-bearing winds off the Pacific Ocean. Although precipitation is variable from year to year, the Owens Valley averages only five to six inches annually and well deserves the title Mary Austin awarded it in 1903: the land of little rain.[10]

The aridity of the Owens Valley has generally tended to limit vegetation to hardy, drought-resistant plants even though the natural vegetation of the region has been significantly altered over time. But subterranean conditions in the Owens Valley, as elsewhere in the intermountain West, have influenced biotic communities.[11] At the base of the alluvial

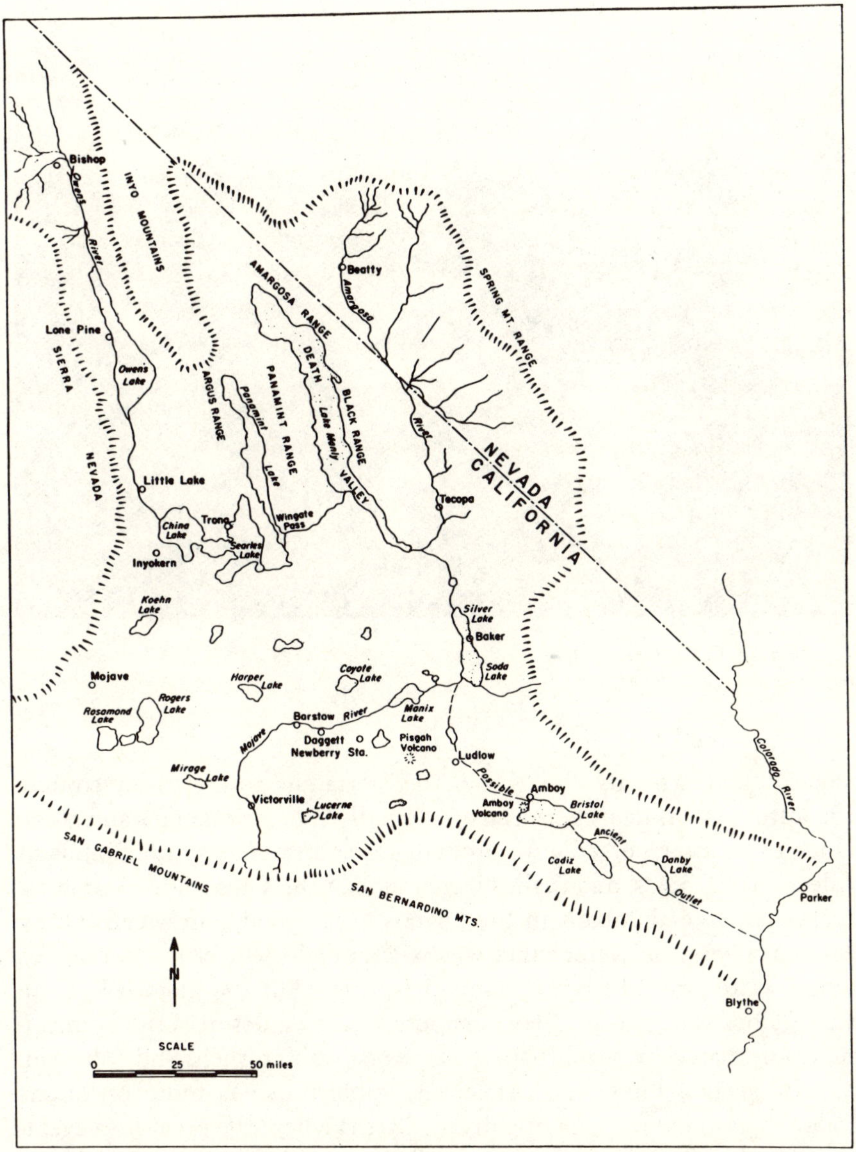

Figure 2.5. The interconnected network of Pleistocene lakes in east central California, of which Owens Lake was a part. (Reproduced from Blackwelder, "Pleistocene Lakes and Drainage," 36)

Figure 2.6. The Owens River.

slopes where streams flow out of the Sierra and in the valley trough, the water table is near the surface, originally causing marshes and grasslands to be interspersed with drought-resistant desert scrub. Stands of tules, cattail, reeds, and rushes were found at these locations. Salt grass meadows were abundant in the Owens River floodplain because they thrived where the water table was within eight feet of the surface or where water spread from streams or springs. But the majority of the valley floor was, and is today, characterized by a desert scrub community dominated by sagebrush, greasewood, rabbitbrush, and saltbush, and by several bunchgrass species.[12] Rabbitbrush is more prevalent today than in the past, having invaded areas where the previous vegetation has been weakened or removed. On the valley's alluvial slopes, sagebrush is the dominant species, while at higher elevations on the mountain slopes, woodland predominates. Of considerable importance to the valley's native subsistence was the open woodland association of juniper and piñon pine found in the White-Inyo Mountains at the 6,500-foot level.

The earth movements that have occurred throughout the Owens Valley's history have affected drainage and the accumulation of soil materials washed down from the mountain slopes.[13] Generally, in the Round Valley–Bishop region, recent alluvial deposits composed of sandy loams and fine sandy loams are found in the floodplain of the Owens River and on level surfaces traversed by the prolific flow of numerous Sierra streams, which formerly spread out over the valley floor.[14] South of Poverty Hills, the Sierra streams are more widely dispersed, and the soils are generally sandy, more poorly watered, and more alkaline. Much of the southern valley, if properly supplied with water and washed of its salts, is suitable for agricultural purposes. Nevertheless, the north end of the valley, from Big Pine to Round Valley, originally contained the best combination of abundant stream flow, lush meadows, and fertile, alkali-free soils, features that were recognized by both the valley's aboriginal inhabitants and its pioneer settlers.

Prehistoric Paiute Ecology

The Owens Valley Paiute comprise the southernmost bands of the Northern Paiute group, which occupied most of northern Nevada and the western margin of the Great Basin. As in other parts of the Great Basin, the relationship of the Owens Valley's native inhabitants to their environment was, as Julian Steward has called it, "simple and direct."[15] Their food supply was derived from a seasonal round of activities that included gathering wild seeds and roots on the valley floor and piñon nuts from the White-Inyo Mountains; hunting deer, antelope, mountain sheep, jackrabbits, and waterfowl; and fishing the Owens River and its tributary streams. Plant foods were more important than meat in Paiute subsistence, lowland roots and seeds were more important than piñon nuts, and small game was more important than large game.[16] Most activities in the annual round occurred in the spring, summer, and fall when the aboriginal population procured, processed, and stored a variety of desert scrub seeds and roots, and piñon nuts from the mountain slopes.[17] To facilitate these activities, the bands subdivided the valley into elongated political units tied to the streams that flowed from the Sierra Nevada. The districts extended from west to east between the summits of the bordering mountain ranges and encompassed various life zones in which each band communally held the hunting, seed gathering, and fishing rights.[18] The districts were led by headmen, who directed the communal activities. Many of the band-owned seed-gathering areas were further subdivided into family plots, as were the piñon nut areas. Rancherias (permanent villages located in the desert scrub community on the valley floor) served as year-round base camps in the seasonal cycle of activity, but some Paiutes also occupied alpine camps to harvest the piñon nut crop in the fall.

To enhance the natural bounty of the land, the Owens Valley Paiute diverted many of the streams flowing off the Sierra to the natural meadows on the edge of the valley. They built temporary dams of boulders, sagebrush, and earth across these streams, then cut shallow ditches to divert water toward the more prolific wild food species, particularly yellow nutgrass. They did not till, plant, or cultivate the soil, "but cleverly watched how nature waters the grasses and bulbs, then followed suit."[19] Julian Steward maintained that the Owens Valley Paiute practiced irrigation without agriculture through "intensifying by irrigation what nature had already provided."[20] Other scholars contend that the Owens Valley Paiute, in expending tremendous effort in constructing the dams, ditches, and canals, and in irrigating and harvesting the crops, had developed a complex system of irrigated vegeculture and therefore by definition were practicing agriculture.[21] Whether it was agriculture or not, it is generally conceded that the practice of irrigation among the Owens Valley Paiute was an indigenous invention.[22]

The irrigated plots were located at ten sites along the west side of the valley and were most numerous in the better-watered north end of the valley (Fig. 2.7). The largest was known as *pitana patü* and was located in two large natural meadows on either side of Bishop Creek (Fig. 2.8).[23] The north side of the creek held the greatest abundance of *tüpüsi'ⁱ* (yellow nut-grass) and *náhavita* (wild hyacinth), the two most important plants harvested from the irrigated plots in the Owens Valley because of the edible nutlike bulbs at the base of their stems. The more widespread tüpüsi'ⁱ came to be known as grass-nuts or "taboose grass" by white settlers and later generations of Indians. The northern plot at pitana patü was about six square miles in extent and was fed by an irrigation ditch two miles long. The southern plot, which also contained both plant species, was perhaps half the size of its northern counterpart and was watered by a ditch three miles in length. The two plots were alternately irrigated annually; in one year, water was turned into one plot in the spring, and the next year the other was irrigated. This form of fallowing presumably was practiced to enable the plots to reseed naturally. It was an honor to hold the position of irrigator at pitana patü. The irrigator was elected at a popular meeting each spring, after which the beginning of the irrigation season was announced by the district headman and approved by the people. The dam was built by the irrigator, who was assisted by about twenty-five men, and the irrigator

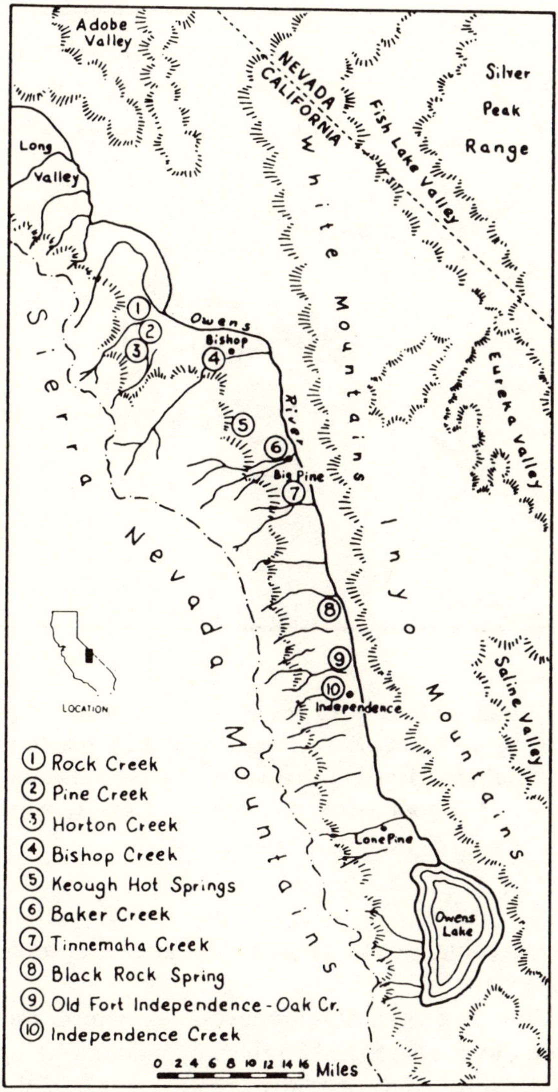

Figure 2.7. The location of irrigated plots of the Owens Valley Paiute. (Reproduced from Lawton et al., "Agriculture Among the Paiute of Owens Valley," by permission of the Malki Museum, Inc.; based on Map 1 in Steward, *Ethnography of the Owens Valley Paiute*)

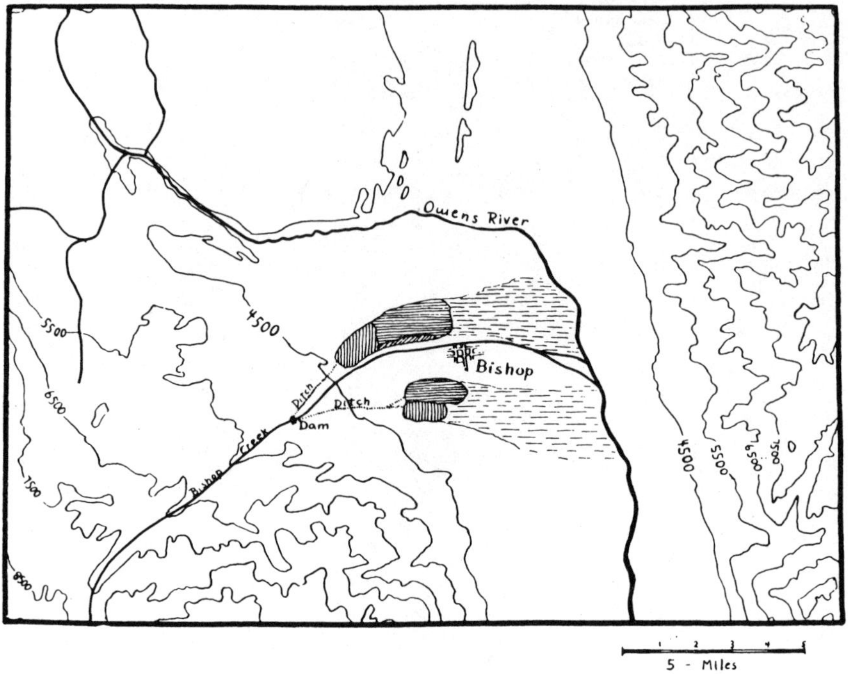

Figure 2.8. Irrigation at *pitana patü* near present-day Bishop. The vertical cross-hatching represents plots of *tüpüsi'*^{i}, and the horizontal cross-hatching is *náhavita.* The dashed lines are natural grassland areas. (Reproduced from Steward, "Irrigation Without Agriculture," 151, by permission of the Michigan Academy of Science, Arts & Letters)

was solely responsible for watering the plot after the water was turned into the ditch.[24]

Seed-yielding plants – including love grass, wheat grass, wild rye, and sunflower – were also found on the valley floor, often growing prolifically on areas watered by overflow from the irrigated plots.[25] Groups of Paiute women gathered seeds in the summer and fall by beating them from the plants and dug the tubers and roots with pointed sticks.[26] The corms of náhavita and tubers of taboose were stored and later roasted and ground into flour. The mountain piñon camps were occupied by small groups in the fall during harvesting. No evidence of irrigation has been found in the southern portion of the valley between present-day Independence and Owens Lake. Natural conditions were somewhat less conducive to irrigation in this region. Here the larvae of the small fly that formerly occurred in the alkaline waters of Owens Lake were gathered in such great abundance by the Owens Lake communities that this resource substituted for irrigated tubers.[27]

Steward conceded that the Owens Valley Paiute were probably on the verge of horticulture at the time white settlement began but that they had not yet quite achieved it.[28] Instead, he considered their irrigation practices simply an artificial reproduction of natural conditions. When lateral streams sometimes overflowed their banks and flooded the land, the Paiutes no doubt observed the expansion of taboose and other plants that otherwise were limited to areas of permanently moist soil. Through irrigation they simply expanded the natural conditions in the valley.[29] The Owens Valley Paiute were therefore the first humans to manipulate and modify the natural ecology of the valley, with only minimal disturbance of the region's ecological balance even though the Owens Valley was apparently one of the most densely settled areas of the Great Basin region in prehistoric times.[30]

Population estimates of the Owens Valley Paiute at the time of white settlement range from 1,000 to 2,000.[31] Most of the Indians lived in conical shelters made of thatched tule or willow reeds clustered into an estimated thirty rancherias between Round Valley and Owens Lake.[32] The unusually varied environment of the Owens Valley and the bordering mountains provided all of the people's essential food resources within twenty miles of the villages.[33] With the adoption of irrigation, the Owens Valley Paiute developed a seasonal round of subsistence activities that harmonized with their arid habitat. In the early 1860s, however, with the onset of white settlement, the ecological relationships that the Paiutes had established in the shadow of the Sierra were permanently disrupted.

3. Early Views of the Land

> For some distance along the river, after leaving the Lake, but little desirable land is found, except that supplied with water by little rivulets flowing from the mountains. Twenty-seven miles from the head of the Lake is Pine Creek [now Oak Creek], with a large body of meadow land and the first timber we encountered growing in the valley, save a few small cottonwoods. Beautiful streams of clear cold water come gushing fresh from the snows of the Sierra, at intervals of from one to ten miles, irrigating beautiful and fertile portions of the valley, for the following sixty-two miles.
>
> – "Quis," in the *Los Angeles Star,* August 27, 1859

Walker's Expeditions

The earliest known expedition into the Owens Valley was led by Joseph Reddeford Walker, who set out in 1833 from the Great Salt Lake region on a beaver trapping expedition to California. On his return journey in the spring of 1834, Walker searched for a path from the San Joaquin Valley through the Sierra Nevada in order to intercept the Humboldt River, the route used in his outward journey across the intermountain West. In late April, guided by local Indians, Walker's party moved up the south fork of the Kern River, crossed the Sierra over the pass ever since known by his name, and descended the east face of the range, reaching its base some distance south of then-unnamed Owens Lake. In early May, as Walker's party traveled north along the eastern flank of the Sierra, Zenas Leonard, the expedition's clerk, wrote: "The country on this side is much inferior to that on the opposite side – the soil being thin and rather sandy, producing but little grass. . . . On the opposite side vegetation had been growing for several weeks – on this side, it has not started yet."[1] On the trek north through the Owens Valley, Leonard did make note of numerous small rivers that flowed out of the mountains and emptied into the desert, "where the water sinks and is exhausted in the earth."[2] At these locations Leonard recorded the presence of natural meadows that provided "tolerable pasture" for the horses of Walker's expedition.[3]

Although Walker led several expeditions through the Owens Valley in the 1830s and 1840s, the valley remained relatively unknown until the 1850s. Walker himself apparently kept no journals. Zenas Leonard's brief and somewhat disjointed description of the Walker party's original trek through the Owens Valley, part of which he wrote from memory because a portion of his narrative was stolen from him by Indians, is the earliest known written record of the region. In 1845, Walker and a companion, Richard Owens, joined another explorer, John C. Frémont, who had set out from Bent's Fort in Colorado on his third expedition to the West.[4] A south-bound party from the expedition, including Frémont and Owens, crossed the Sierra before reaching the Owens Valley by traveling up the Truckee River. W. A. Chalfant, the Owens Valley's official chronicler, claims that Frémont was so impressed with Owens's capabilities as a mountaineer and hunter that he later gave his name to the eastern Sierra lake from which the river and valley took their name, even though Owens never traversed the region that today bears his name.[5] The main expedition, which included Walker, traveled south through the Owens Valley and then crossed the Sierra through Walker Pass to rejoin Frémont. Edward M. Kern, a photographer and topographer, kept a diary of the journey, though his description of the valley is nearly as brief as Leonard's.[6] He did, however, include two important observations regarding the valley's natural conditions. As the party traversed the central portion of the valley, Kern noted that they were "obliged to keep some distance from [the Owens River] on account of a large marsh."[7] He also described how Walker, when he went in search of some salt, found it "incrusted to the thickness of a quarter of an inch on the surface of the earth."[8] Both the extensive tule marsh and the alkali accumulations that Kern recorded would discourage later settlers.

The Federal Land Survey

The first detailed account of the Owens Valley was provided in 1855 and 1856 by A. W. Von Schmidt, who, under federal contract, conducted the first public land survey of the Owens Valley. After the Mexican land cession of 1848, much of the Far West, including the Owens Valley, was incorporated into the public domain, and the public land surveys that followed were the initial step in the government's securing of effective control of these new lands.[9] In accordance with federal instructions,[10]

Von Schmidt systematically evaluated the resources of the valley, recording the character of the terrain, the quality of the soil, the nature of the vegetation, and the prevailing water supply. Initially Von Schmidt surveyed only the township lines, working from north to south through the valley. In July 1855 he summarized his impressions:

> The Valley that I worked the line down will average about 6 miles in width and bears off in a S. E. by S. direction. The land in this Valley is entirely worthless with but few exceptions the only portion of any value is near the banks of the little streams of water coming from the Sierra Nevada Mountains.
>
> On a general average the country forming Owens Valley is worthless to the White Man.
>
> Both in soil and climate.[11]

No white settlement then existed in the valley. With no land under cultivation and only a scattering of trees along the stream courses, the Owens Valley's sagebrush-covered stretches were apparently, as Chalfont notes, "not inviting to one fresh from the springtime aspect of the grassy and flowered hills of western California."[12]

Von Schmidt's subsequent northward trek to survey the townships' section lines in the late summer and fall of 1856 produced a more detailed and somewhat more positive description of the region. The survey began in mid-July, but after surveying only five townships Von Schmidt temporarily halted his work in early August out of fear of the Indians, who had killed several of his mules.[13] After increasing the size of his survey party, he resumed work in mid-September. To hasten the project, he ran two survey teams northward through the valley, leading one himself and naming Deputy Surveyor E. K. Whiting to lead the other. By the time the survey was again underway, the irrigated plots of the Owens Valley Paiute were near harvest and looked different from the previous summer. The Indian irrigation of " 'sabouse' [taboose] . . . their principal article of food," which Von Schmidt recorded in his field notes, no doubt caused him to reconsider his initial views.[14]

Because of the valley's narrowness, most townships were only fractionally surveyed, with the survey discontinued at the base of the mountains or where the alluvial slopes became too steep and rugged. Von Schmidt categorized the majority of the valley's soils as second- or third-rate; some, however, he described as first-rate (Fig. 3.1). A review of Von Schmidt's field notes reveals a close correlation between the prevalence of well-watered areas of "fine grass" and first-rate soils. The largest expanse of first-rate soils, according to Von Schmidt, was located

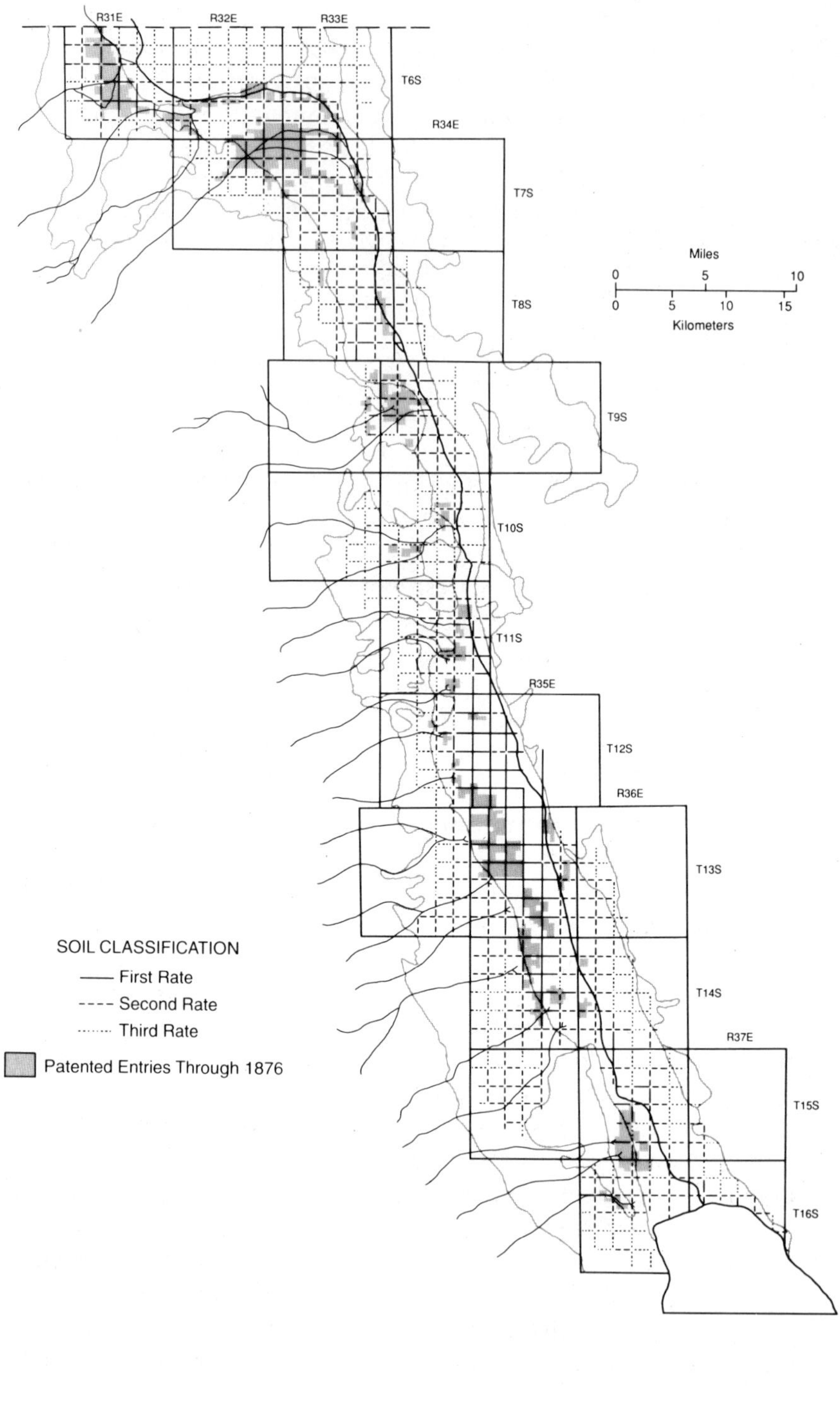
R31E
R32E
R33E
R34E
R35E
R36E
R37E
T6S
T7S
T8S
T9S
T10S
T11S
T12S
T13S
T14S
T15S
T16S
Miles
0
5
10
0
5
10
15
Kilometers
SOIL CLASSIFICATION
First Rate
Second Rate
Third Rate
Patented Entries Through 1876

in T13S R35E, in the vicinity of present-day Independence. Working from Owens Lake northward, this is the first extensive meadow that he would have encountered. But this meadow consisted of a carpet of salt grass growing on alkali-prone land. This was the "large marsh" identified by Kern in 1845. There were instances where well-watered areas of "fine grass" were described as second-rate soils. This discrepancy was probably due to differences in opinion between the two survey teams. Obviously, the methods used to classify soils in those days were more subjective than scientific; the *relative* quality of natural vegetation had a significant influence on soil ratings. On the opposite side of the Sierra in the Tulare Lake Basin, for example, federal land surveyors described timbered areas as having first-rate soils while designating grasslands as second- or third-rate.[15] Only the grass-covered areas of the Owens Valley impressed Von Schmidt; nowhere in his field notes is there any mention of the widespread occurrence of sagebrush and other desert scrub vegetation. One must assume, therefore, that except for the valley's natural meadowlands, Von Schmidt believed the valley to be worthless for agricultural purposes.

Capt. Davidson's Impressions

Three years after Von Schmidt's survey, in July 1859, a punitive expedition led by Capt. J. W. Davidson was sent from Fort Tejon to the Owens Valley to investigate charges that the local Indians had been rustling horses from other regions of California. When it quickly became apparent that the Owens Valley Paiute were not responsible for the stolen horses, Davidson turned his attention to a careful examination of the land and life in the Owens Valley, for he had also been ordered to "examine the country well with reference to its fitness for the purpose of an Indian Reservation" and to survey its agricultural, timber, and water resources.[16]

Davidson's subsequent account – entitled "Report of the Results of an Expedition to Owen's Lake, and River, with the Topographical Features of the Country, the Climate, Soil, Timber, Water, and also, the

Figure 3.1. Soil classification in the Owens Valley, based on the original federal land survey, and patented entries through 1876. Lines on the map identify streams and landforms shown in Figure 2.1. (Compiled from Von Schmidt, Field Notes, and Independence District Land Office tract books; redrawn from Sauder, "Sod Land Versus Sagebrush," 408)

Habits, Arms, and Means of Subsistence of the Indian Tribes seen upon the March" – represents one of the most thorough descriptions of the Owens Valley prior to settlement.[17] In fact, Davidson's impressions, a popular version of which appeared in the *Los Angeles Star* in late August 1859, were so glowing that they may have been partly responsible for the movement of stock raisers into the region during the three-year drought that affected the rangelands in the San Joaquin Valley and southern California beginning in 1861.[18]

Descending into the San Joaquin Valley from Fort Tejon, Davidson traveled over the Sierra via Walker Basin to the south fork of the Kern River and then followed Walker's original route over Walker Pass. Moving north, he skirted the western shore of Owens Lake and journeyed up the west side of the Owens Valley to Round Valley and the Owens River Gorge. As Davidson traversed the valley from south to north he recorded his overall impression: "Every step now taken shows you that nature has been lavish of her stores. The Mountains are filled with timber, the vallies with water, and meadows of luxuriant grass. Some of the Meadows contain, at a moderate estimate, ten thousand acres, every foot of which can be irrigated." The area north of Independence, in T13S R35E, "where the eye wanders over a sea of green," impressed Davidson as much as it had Von Schmidt. So did Round Valley, "the finest and most picturesque we have yet seen," where in one meadow "the grass (over two feet in height, broad leaved, and juicy) extended for miles."[19] After completing his mission, Davidson concluded that he had marched through some of the finest country he had ever seen: "a vast meadow, watered every few miles with clear, cold mountain streams, and the grass (although in August) as green as in the first of spring."[20]

As with Von Schmidt, the valley's sod lands were the focus of Davidson's attention, while the more widespread expanses of sagebrush were entirely neglected in his report. Davidson did record the existence of abundant stands of "pine timber of fine quality and easy access" at Bishop Creek, Big Pine Creek, and Independence Creek, estimating that they would provide enough lumber "for all the uses of a population commensurate with the agricultural resources of the valley."[21] Davidson also found a grove of "magnificent oaks" growing along Oak Creek. Because of the ubiquitous snow-fed streams flowing off the Sierra, Davidson believed that the Owens Valley was "the finest watered portion of the lower half of the state," and that "wherever water touches [the soil], it produces abundantly." Assessing the region's agricultural potential, Davidson speculated that the Owens Valley was well-suited to the production of wheat, barley, oats, rye, and fruit such as apples and pears: "To the Grazier, this is one of the finest parts of the state; to the Farmer, it offers every advantage but a market."[22] Davidson's observations regarding the agricultural possibilities in the valley would prove to be accurate.

The California Geological Survey

William Brewer, who led field parties for the California Geological Survey under the direction of Josiah Whitney in the early 1860s, provided a brief natural description of the Owens Valley just as permanent white settlement was beginning. In the summer of 1864 he at least alluded to the widespread occurrence of desert scrub when he recorded the following: "The aspect of these deserts is peculiar. In the distance, when individual bushes cannot be distinguished, they look like a gray plain of uncovered soil; near by they are still gray, or yellowish gray, but covered with bushes." Three days later Brewer wrote: "We went up the valley sixteen or seventeen miles where a lava table crosses the entire valley, barren in the extreme. Here a stream comes down, and there is a sort of basin where there are nine or ten square miles of the best grass I have seen in the state."[23] Brewer was referring to the natural meadows found in the valley's northern townships around present-day Bishop.

It is evident that those who entered the valley prior to permanent settlement and recorded their impressions were particularly impressed by the lush green meadows found in the region. These grass-covered expanses occurred naturally in the best-watered sections of the valley, but their areal extent had been somewhat enlarged by the valley's native inhabitants. After the Indians had been banished from the valley with the onset of permanent settlement, their old irrigated fields were the first to be taken up by the new arrivals. The *Inyo Independent,* in its second issue, published on July 13, 1870, predicted that "in the possession of the whites, they [the patches of meadowland] will soon be converted into luxuriant gardens, orchards, and grain fields."

4. Thoroughfare to the Mines

I repeated to them what I had said at Dragoon Fork, that their treatment by our government depended upon their good or bad conduct; that the road through their Country must be an open one, that I had reason to believe that their Country was set apart by the Government – exempt from settlement – for their use so long as they maintained honest and peaceful habits; but that if they ever changed this manner of life, for one of robbing and molesting the white people, they would assuredly be punished by the Sword.

– Capt. J. W. Davidson, 1859

The whole of Owen's Valley, with the different mining camps in that vicinity . . . is entirely exposed to the attacks of the Indians, and within sixty or eighty miles of Owen's Lake there is an immigration of about fifty large wagons (going to Aurora, Mono County) loaded with valuable goods and machinery which can reach their destination by no other road than through Owen's Valley. Besides this there are also on the road a great many thousand head of cattle, sheep, and hogs for the same destination.

– W. L. Kennedy, April 23, 1862

A Market in the Mines

Because the Owens Valley was distant from immigrant routes into California, land and life in the region remained largely unaffected by the California Gold Rush of 1849. But ten years later, when the mines of the western Sièrra Nevada began to play out, prospectors crossed the Sierra in search of new veins of ore in the western Great Basin. Gold and silver discoveries at Monoville in 1859 and Aurora in 1860 quickly transformed the Owens Valley into a thoroughfare to the boom towns of the eastern Sierra.[1] Aurora became the more important of the two mining centers, in the early 1860s rivaling Virginia City as the West's most spectacular mining town.[2] Miners swarmed into Aurora during the winter of 1860–61, and by the summer of 1861 it had nearly two thousand inhabitants. The town became the outfitting center for prospectors who roamed the western Great Basin, and it continued to grow through the summer of 1863, when its population reached about five thousand.

Most of the supplies for this expanding population were freighted in from either Sacramento or Visalia, but stock raisers from the southern San Joaquin Valley and the Tejon country of southern California also responded to the demand for food by launching, in the spring of 1861, the first cattle drives to the new mining centers. As they were trailing their cattle over Walker Pass and into the Owens Valley, the stockmen could not fail to notice the fine grazing lands of the region. Further, they quickly realized that the long drive could be avoided by establishing ranches in the valley, the northern portion of which was only sixty-five miles from Aurora. In August 1861, Samuel Bishop established the first stock ranch in the valley when he drove between 500 and 600 head of cattle from Fort Tejon to its northern end.[3] The location of Bishop's ranch, about three miles southwest of the town that today bears his name, "at a point where the stream leaves the higher sandy bench lands and gravel foothill slopes and enters the lower level of the valley,"[4] placed it close to, if not on, the Paiute's irrigated fields at *pitana patü*.

White Encroachment on Paiute Lands

By that fall, nearly a dozen stock raisers had established ranches in the valley.[5] Cattle grazing on the natural meadows resulted in the destruction of native plants, whose seeds and roots were the staples of the Paiute diet. A correspondent for the *San Francisco Evening Bulletin* who trekked through the Owens Valley in June 1861 described the fondness of cattle for plants found precariously near the irrigated fields of the Paiutes: "Most of these streams . . . [are] turned aside for irrigation, a purpose to which they are extensively applied by the Indians. The surplus water flowing over the land below these patches of roots has caused much grass to grow along these creeks, consisting of clover, blue-joint, and bunch grass. Cattle are very fond of these and fatten upon them rapidly."[6]

The mining activities of the region also impacted the livelihood of the Owens Valley Paiute. Lumbering in the adjacent mountains, which occurred as an adjunct to mining developments, began to deplete the piñon forests upon which the Paiute depended as a source of winter food. During the particularly severe winter of 1861–62, when the cutting of piñon pines for the mines and the grazing of livestock on the irrigated meadows impacted their winter food supply, the Paiute were forced to prey on the cattle that grazed in the valley.[7]

Two years earlier, Captain Davidson had discovered an awareness among the native population of the valley that they would eventually have to come to terms with the outside forces. Von Schmidt's survey of the valley, combined with the migration of prospectors through the region and Davidson's own expedition, no doubt led the Indians to this conclusion. Davidson, realizing that "squatters, when the resources of this country become known, must soon swarm over their lands," recommended that immediate steps be taken to establish a reservation encompassing the entire valley "before the Indians shall have learned the vices which the white man brings in his train."[8]

In addition to the cattle herds of the growing number of permanent ranches in the Owens Valley in the early 1860s, ranchers drove thousands of head of cattle that were starving (because of the prolonged drought affecting western California at the time) over the Sierra Nevada to feast on the valley's luxuriant summer pastures. The increased number of cattle exacerbated the problem, resulting in stepped-up Indian attacks on whites, many of whom were driven out of the valley. Those who remained sought protection by requesting that a military presence be established in the valley. Colonel James W. Carleton, the commanding officer of the First California Volunteers at district headquarters in Los Angeles, however, was at first not convinced of the necessity of locating a military post there:

> I have heard that the white men went into Owen's Lake Valley and took their stock onto the fields of grass which is cultivated by those Indians, the root of the grass being used by them as an article of food, and that, on the Indians remonstrating with them, they persisted in keeping their stock in these fields. The Indians of Owen's Lake Valley are known to be an inoffensive, gentle race. Up to this time they have never been hostile to the whites. They live mostly on the larvae of a fly which lays its eggs upon the water of the lake. The Indians there depend on this article and on the grass roots for their subsistence. So you can judge that a people who require the use of no other weapons than such as are necessary to procure these supplies are not very dangerous as an enemy. It was the intention of the Government at one time to make that region of country a reservation for all the Indians in Southern California. Since the Washoe and Coso mines have been discovered, people have gone into this valley with stock, to be convenient to the mines for the sale of beef, &c., and the poor Indians are doubtless at a loss to know how to live, having their fields turned into pastures

whether they are willing or not willing. It is very possible, therefore, that the whites are to blame, and it is also probable that in strict justice they should be compelled to move away and leave the valley to its rightful owners. Another consideration must be taken into account. The whites there may desire to have a military post in that vicinity, as well to protect them as to be a market for their beef. The question will be as to the necessity for such a post.[9]

In spite of Carleton's initial assessment of the Indian situation, the following month he sent Colonel George S. Evans into the valley to take a closer look. Evans reported:

From all the information that I could get and from what I know of my own knowledge, I am of the opinion that the Owen's River Indians, together with detachments from the Tejon, Tulare, and Mono Indians, and some of the Piutes, have banded together, numbering not far from 800 to 1,000 warriors; that they have 100 or more good guns, and are determined to carry out their threat that no white man should live in the valley. They have killed at least 1,000 head of cattle, and have been drying the meat and preparing evidently for a long war, and to-day there is not a white soul left living in the valley. The mining interests of that section are too great for the whites to give it up tamely. Again, the Owen's River Valley is the great thoroughfare and only route, except to go around by Placerville, through which the growing trade and travel of this southern country must pass in and to the Esmeralda and Washoe districts, and upon which the people of Esmeralda are almost entirely dependent for their beef and other fresh meats. In consideration of these facts and in compliance with my instructions, I would most respectfully urge the necessity of a military post being established in the valley.[10]

Evans's argument was persuasive, and in late June 1862 he was sent back into the Owens Valley with orders to locate a garrison. As the Colonel scouted the valley for a suitable site for the post, he noted the following:

The Indians claim the valley as belonging to them, and still insist upon it that no white man shall settle, or, as they term it, sit down in the valley. They say that the whites may pass through to and from Aurora if they want to, or they may locate in the hills and work the mines, but must not sit down on the grass patches. Now, without arguing the point as to their right by prior location to the exclusive use of the valley, I will say that it is very evident to my mind that the mines will be of small value unless the valley can be settled and grain and vegetables grown and beef raised to feed the miners with.[11]

Evans concluded that the best way to subdue the native population was to locate the post somewhere near the center of the valley, from which scouts could continually range from one end to the other, destroying winter supplies of grass seeds and fly larvae and preventing the Paiutes from gathering additional supplies. "Without this food gathered and laid up," wrote Evans, "they cannot possibly subsist through the winter."[12] In a treeless country where the Indians could observe the movement of troops miles in the distance, Evans argued that it would be easier to starve them into submission than to fight them. On July 4, 1862, Evans located a site for the garrison on Oak Creek about two miles north of the present town of Independence and named it Camp Independence in honor of the day.

Within a few months, Evans's strategy seemed to have succeeded; the first truce was signed between the military and the Owens Valley Paiute in October 1862.[13] The agreement, however, was short-lived. It was based on the recommendation of the superintendent of Indian affairs for the Southern District of California that a reservation be established in the valley setting aside six townships located between Big Pine Creek on the north and George's Creek on the south – comprising nearly 140,000 acres – for the estimated two thousand Indians who resided in the valley. It was noted that only in small areas was the land capable of cultivation, "the balance being scarcely fit for grazing purposes, and none of it attractive to settlers."[14] But the proposed reservation was disallowed by the commissioner of Indian affairs, and even if it had been approved, warfare probably would have continued. For the Paiutes to maintain their seasonal round of activities, it was essential for them to continue to irrigate, gather, and hunt throughout the region's different life zones.

Failure of the agreement meant that Indian hostilities continued during the winter of 1862–63, but the reduction of the Indians' food supplies caused many of them to surrender at Camp Independence in June 1863. The following month more than nine hundred Paiutes were removed to the San Sebastian Reservation near Fort Tejon, and with Indian warfare now seemingly over, Camp Independence was ordered disbanded. Most of the Indians, however, gradually slipped away from the reservation and returned to the valley. Sporadic Indian attacks against whites resumed, making travel through the valley dangerous once again. In the spring of 1865 Camp Independence was reoccupied, but only after the ruthless massacre of forty-one Paiutes on the shores of Owens Lake at the hands of a white militia. Relatively few clashes occurred subsequent to the re-establishment of the garrison in the valley.

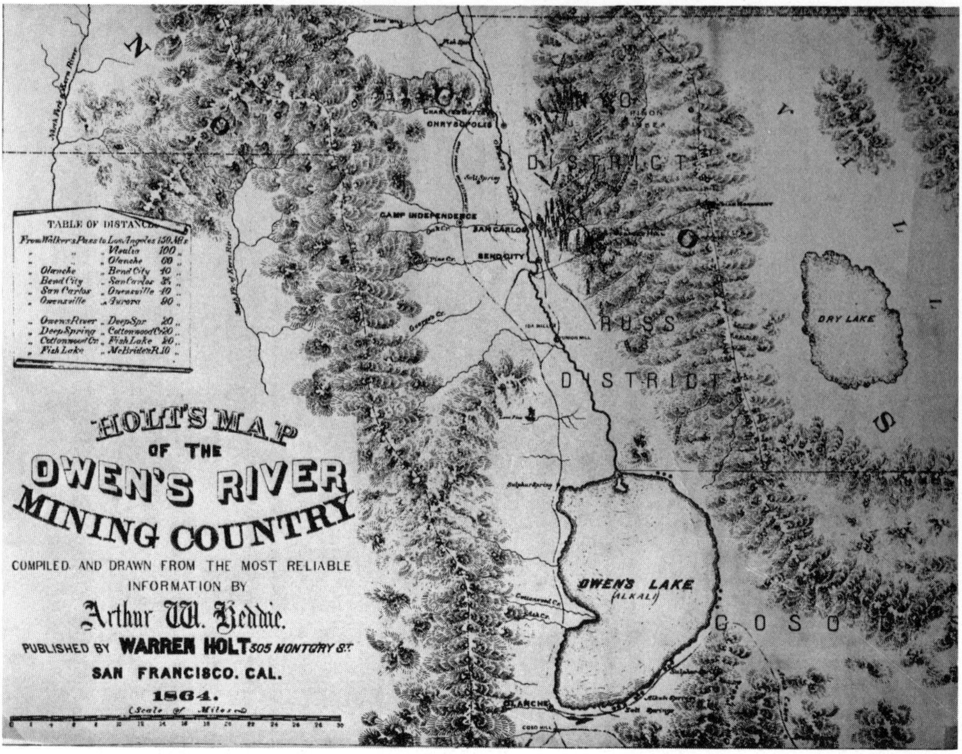

Figure 4.1. Detail of *Holt's Map of the Owen's River Mining Country*, showing early mining districts in the vicinity of the Owens Valley. (San Francisco: Warren Holt, 1864; courtesy of the County of Inyo, Eastern California Museum, Independence)

In all, the Owens Valley conflict cost the lives of at least two hundred Indians and thirty whites.[15]

The Formation of Civil Government

Clashes between whites and the Owens Valley Paiute retarded the progress of mining activity in the region. With the reestablishment of Camp Independence, however, prospecting resumed, and several mining centers emerged along the east side of the valley in proximity to the mines. Owensville was the chief mining camp at the north end of the valley, serving the mines located on the western slope of the White

Mountains. Farther south, facing the Inyo Range, were the towns of Chrysopolis, San Carlos, and Bend City (Fig. 4.1). All reached their peak between 1864 and 1866, and each was subsequently abandoned in response to the rise of more permanent farming communities on the valley's west side. In July 1864, William Brewer recorded the lethargy that prevailed in one of these mining communities:

> We camped on the river near Bend City and went into town for fresh meat and to get horses shod. It is a miserable hole of perhaps twenty or twenty-five adobe houses, built on the sand in the midst of the sagebrush, but there is a large city laid out – on paper.
>
> It was intensely hot, there appeared to be nothing done, times dull, and everybody talking about the probable uprising of the Indians.[16]

Times remained dull in the Owens Valley until the discovery of the rich Cerro Gordo silver deposits east of Owens Lake in 1865. Mining activity in Aurora was now on the wane, and interest had shifted south. The Cerro Gordo discovery injected significant economic life into the region, and a more permanent class of settlers began to arrive in the valley in the mid-1860s. Scattered homes appeared up and down the length of the valley, and the newly arrived population began to push for the formation of local civil government. Although the creation of mining districts in the early 1860s was the harbinger of local government in the region, most of the Owens Valley still lay within the "shadowy jurisdiction" of Tulare County.[17] Subsequently, the residents of the Owens Valley memorialized the California legislature for the formation of Inyo County, and approval came in March 1866. The largest chunk of the new county was carved from Tulare County, which willingly parted with the remote eastern Sierra region, and the new town of Independence was selected to be the county seat.[18] Thus the whites' discovery of the resources contained in the mountain slopes of the western Great Basin and the onset of mining in the Owens Valley region led to the subjugation of the Paiutes by military force and the dispossession of their lands, to the establishment of an economic motive for the agricultural colonization of this arid and isolated valley, and to the formation of local civil government. The same process was repeated on other frontiers scattered throughout the intermountain West.[19]

5. The Process of Pioneer Settlement, 1860–1880

Some of the prospectors left the country, and many of the miners followed, but the wise ones had seen farther into the future of this land, and gave up their picks and drills for the spade and plow. Ranches were taken up, irrigating ditches run out from the mountain streams that came tumbling down from the Sierras, and in a short time farming communities began to spring up between the Owens river and the high Sierras.

– "In the Land of Plenty" *Inyo Magazine*, August 15, 1908

Federal Land Alienation Measures

Pioneering in North America took settlers into a succession of new and untried environments, and generally their success, or failure, was based on simple trial and error.[1] Initially, this often meant locating vegetation and other surface features that resembled those known to be productive in areas from which the pioneers had migrated. That is, when the land in new environments looked right, it was taken. But when the American frontier advanced from the timber and grasslands of the humid East into the dry, sagebrush-covered lands of the intermountain West, it was apparent that a new set of adjustments would be required for successful settlement.

The most immediate concern of pioneer settlers in the Owens Valley was how to colonize a region so isolated and so different from the areas from which they had come. With the completion of the federal survey of the valley in the mid-1850s and the expansion of mining activities beyond California's Mother Lode into the western Great Basin in the late-1850s, the demand for farm products in the region increased, and the stage was set for pioneers to improve the rectangular units that they had selected for settlement. The transfer, or alienation, of the Owens Valley's public lands to private ownership generally involved the use of measures that had been designed for settling the more humid regions farther east. All of the significant nineteenth-century land laws, including the Preemption Act, the Homestead Act, the Timber Culture Act, the Desert Land Act, and the state land grants were employed to alienate public land in the Owens Valley. Indeed, throughout the period of land disposal, public land in the valley remained in the unoffered class, meaning that private entry through outright cash purchase from the federal government was prohibited. Instead, specific conditions were imposed on nearly all types of entry.

During the valley's early settlement period, preemption and homesteading were the primary methods of land entry by pioneer settlers, other than the purchase of state school lands. The Preemption Act of 1841 gave settlers a preferential right to purchase a maximum of 160 acres that they had settled on and improved.[2] Although the specific provisions of the act changed over time, during most of the Owens Valley's settlement history, preemptors had eighteen months after filing their declaratory statement, or intention to preempt, to construct a dwelling, begin cultivation, and pay the standard price of $1.25 per acre. In the early 1870s, large quantities of agricultural college scrip were marketed, and many preemption claims in the Owens Valley were purchased with college scrip. The amended Agricultural College Act of 1870 encouraged the use of such scrip by preemptors because its depreciated value allowed them to purchase quarter sections at a price considerably below $1.25 per acre.[3] Only final preemption entries were recorded at the Independence District Land Office; actual settlement had occurred several months earlier.

Under the Homestead Act of 1862, claimants had up to five years (often extended to seven years) to prove up and take title to their land free of charge except for filing fees. After six months of continuous residence, a homestead could be commuted to cash entry with the payment of $1.25 per acre. Homesteaders were also limited to a 160-acre quarter-section. The Homestead Act, like the Preemption Act, was specifically designed to establish small family farms on the public domain. Those who desired additional land could legally acquire 320 acres by filing under both the Preemption and Homestead acts.[4] Since both land measures required proof of residence, however, the first parcel had to be carried to patent prior to entry on the second.

The Timber Culture Act of 1873 entitled an applicant to 160 acres of land naturally devoid of timber if, on the day of final proof eight years after entry, at least 675 trees were in living and thriving condition on ten acres of land. Settlement was not a requirement for patenting timber-culture entries. The stated purpose of the measure was to increase tree cover in subhumid environments in order to provide a much needed supply of lumber and fuel while simultaneously modifying the climate.[5] Only one tree claim was allowed per 640-acre section, thus accounting in part for the comparatively insignificant amount of land alienated under this land measure in the Owens Valley. But perhaps a more important reason why the Timber Culture Act was the least used public land measure in the valley relates to the planting and nurturing of so many trees, a difficult task in this arid, isolated region where subsistence was the foremost concern of the majority of pioneer farmers.

The Desert Land Act of 1877 was the last of the significant nineteenth-century land measures to receive congressional approval and the only one designed specifically to encourage settlement on arid lands, though it included no residency requirement. Congress believed that the need for farm products in the mining states of the West could only be met through the irrigation of arid lands and that larger parcels than could be obtained under the Homestead and Preemption acts were required in order to make their reclamation feasible.[6] Hence, each desert-land applicant was entitled to claim 640 acres. A deposit of twenty-five cents per acre was to be paid upon filing a desert-land claim, and the balance of one dollar per acre was to be paid within three years when proof of reclamation by artificial irrigation was made.

State land grants from the federal government for the support of public education and internal improvements were also used to alienate public land in the Owens Valley. Throughout the public domain, sections sixteen and thirty-six in each township were granted to the states to further their educational programs. If for some reason "school sections" in certain townships were not available, the state could select "lieu lands" in their place. California generally made its lieu selections only after it had received an application by individuals to purchase state lands, its avowed goal being to use its federal land grants to make land available to actual settlers.[7] Land warrants for 160 and 320 acres were sold to individuals who were limited to the purchase of no more than 640 acres and who were required to file affidavits that the land was intended for their permanent residence.[8] Corruption in the state land office, however, too often impaired the intended execution of these policies.[9]

Swampland in an Arid Frontier

Sometimes the operation of well-meaning but poorly conceived land measures offset the advantages of those laws that benefited the pioneer farmer. One blatant incongruity in the growing body of public land law is exemplified by the attempted "swampland" grab in the arid Owens Valley in the early 1870s. In accordance with the Swampland Act of 1850, public lands designated as swamp and overflowed land could be segregated and donated to the state, and the proceeds from their sale were to be used for their drainage. In California, where outright purchase of federal land was prohibited, this land measure was used by some of the leading speculators in the state to acquire extensive acreage, particularly in the Sacramento and San Joaquin valleys.[10] Thus its attempted use in the Owens Valley was no exception.

In 1873, pioneer farmer Josiah Earl was appointed the first register of the newly created Independence District Land Office. Immediately after his appointment, he requested the segregation of about one-third of the valley floor, much of it bordering the Owens River, as swamp and overflowed land (Fig. 5.1). It was common in California for the state's agents to select land not shown by the federal survey plats to be swamp or overflowed land and to permit entries upon these land selections even where settlers had already located.[11] When the original federal survey of the Owens Valley was completed, Von Schmidt had identified portions of only two townships, T12S R34E and T12S R35E, as swampland.[12] In fact, much of the land in the Owens Valley to be segregated under the Swampland Act was actually covered with sagebrush, and at least 94 of the 221 sections that were to be segregated were already occupied by bona fide settlers.[13]

Homesteaders and preemptors who had not yet received title to lands that were to be segregated would have lost them to the state had the swampland applications been approved. For those who already held a federal patent to their land, the register of the land office recommended that the patents be relinquished and that their farms be purchased from the state, thereby preserving their preemption and home-

Figure 5.1. Sections included in the attempted "swampland" segregation in the Owens Valley. (Compiled from information in "A Stupendous Grab," *Inyo Independent*, October 4, 1873, p. 2; reproduced from Sauder, "Patenting An Arid Frontier," 550, by permission of the Association of American Geographers)

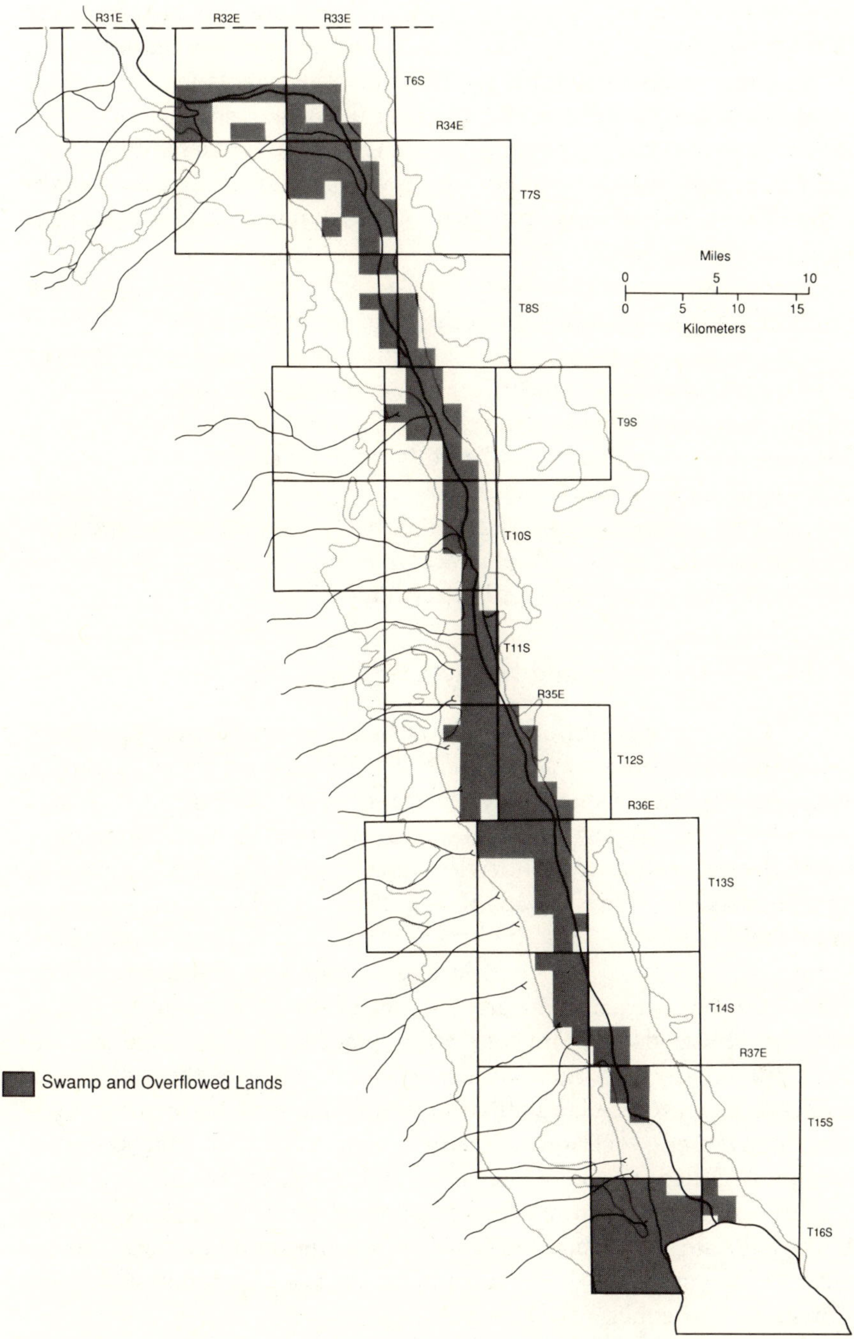
R31E
R32E
R33E
R34E
R35E
R36E
R37E
T6S
T7S
T8S
T9S
T10S
T11S
T12S
T13S
T14S
T15S
T16S
Miles
0
5
10
0
5
10
15
Kilometers
Swamp and Overflowed Lands

stead privileges for a future time.[14] As elsewhere in California, the swampland segregation was made on behalf of land speculators, who, through the process of holding facilitated by the state land office, could retain their selections at very little cost until they could be sold at a profit. Most of the riparian lands of the Owens River were included in the swampland segregation, and the ownership of the stream's banks would have meant control, primarily by stock raisers, of a much larger acreage in the valley.

Upon learning of the swampland scheme, the *Inyo Independent* mounted a vigorous campaign against the attempted grab. "So far, our county has escaped the chief curse of the State," it reported, "but the grabbers are aboard – right in our own midst – and unless we as a people exert ourselves to choke them off . . . our days of freedom from this undeniable curse are numbered."[15] Further action on the swampland applications was subsequently suspended pending an investigation, and they were finally rejected.[16] After the failure of the land grab, the Owens Valley remained relatively free from large-scale speculative activity during its initial period of settlement.

Early Preemption and Homesteading

Prior to the approval of the Timber Culture and Desert Land acts, preemption and homesteading were the primary means of acquiring public land in the Owens Valley. Preemption and homestead entries in the valley peaked in 1873, then gradually declined through the remainder of the 1870s (Fig. 5.2). Initially, pioneer settlers were required to travel across the Sierra Nevada to the district land office in Visalia to enter their claims. In 1869, however, a new land office was opened at Aurora, and it gained jurisdiction over the Owens Valley townships. Since Aurora was then the nearest source of supplies for Owens Valley pioneers, it was also a more convenient place for making land applications. But in 1873, as the volume of land activity in the valley continued to increase, the land office was removed from Aurora to an even more convenient location – the Inyo county seat at Independence. Settlers who had delayed filing their claims because of the inconvenience of traveling to Aurora filed soon after the new land office opened. After that date, the general decline of preemption and homestead entries was associated with the limited amount of unappropriated water remaining in the Sierra streams.

One measure of the success of a land act in fulfilling its desired objectives is the ratio of patented to canceled entries. Successful home-

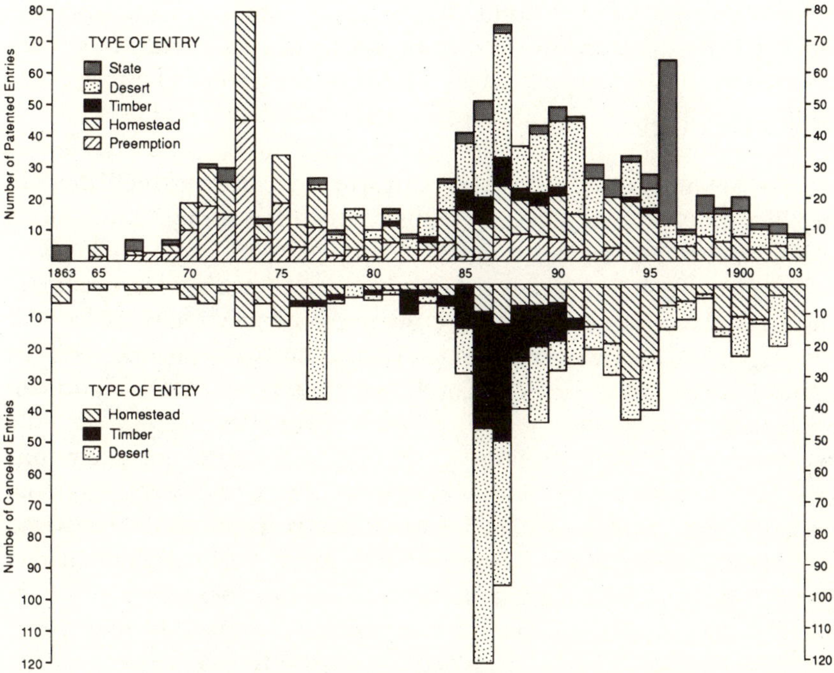

Figure 5.2. The number of patented and canceled entries in the Owens Valley by year of entry and land act, 1863–1903. (Compiled from Independence District Land Office tract books; reproduced from Sauder, "Patenting An Arid Frontier," 553, by permission of the Association of American Geographers)

steading began slowly in the Owens Valley, and during the 1860s the number of claims that were ultimately canceled exceeded those that were carried to patent (Fig. 5.2).[17] This was not necessarily due to inadequacies in the land measure itself but rather to the extreme hardships pioneer settlers initially encountered in this arid frontier.

Eventually, the number of patented homestead entries began to outpace canceled entries. The increased demand for farm products in the nearby mining communities was followed by the growth of the valley's population, and many of the difficulties experienced by the earliest pioneers were gradually alleviated. In the valley's well-watered northern townships, more than 60 percent of the original homestead entries were patented.[18] In the drier southern half, 49 percent of the homestead entries reached the final patent stage. The relatively high proportion of patented entries, particularly in the north end of the valley, demonstrates that the majority of early homesteaders and preemptors were bona fide settlers, just as the authors of the two land laws had intended.

Pioneer Land Selection

Unlike the first white settlers of the Great Basin – the highly organized Mormons – who recorded and preserved their initial observations of their new environment,[19] Owens Valley pioneers generally kept no written records. In order to reconstruct the process of settlement in the Owens Valley, one must examine what the pioneers actually did by correlating the location of initial land selections with the original land cover and other surface features of the valley. Although early reports on the land by Von Schmidt, Davidson, and Brewer may have played a role in settlers' overall impressions of the region, specific land selections were no doubt guided by on-site inspection of surface characteristics by the settlers themselves. Early patented land entries in the Owens Valley therefore represent the first legal recording of the settlement process.[20] Since preemption and homestead applications were generally the only means then available for acquiring public land in the Owens Valley, these early claims, if carried out faithfully, not only prove that settlers believed they could earn a living from the land but also identify the areas the settlers perceived as holding the greatest potential for permanent occupance.

Although the first land locations of record for the Owens Valley were entered in 1863, it was not until 1868 that enough patented entries had been filed to provide early evidence of land capability (Fig. 5.3). The heaviest initial focus of settlement was in the Independence region, no doubt influenced by the location of Camp Independence. In addition to the security that the camp provided, its contingent of a hundred soldiers, as Colonel Carleton had predicted in 1862, represented a convenient market for surplus farm produce. Von Schmidt had also identified much of this region as having first-rate soils. Other early land entries appeared in isolated locations south of the Independence cluster and at Fish Springs, Big Pine, Bishop, and Round Valley in the northern half of the valley. By the end of the decade, the valley's four oldest townsites – Lone Pine, Independence, Big Pine, and Bishop – had been filed upon, and each community grew in response to the addition of new settlers on the fertile land surrounding them. As William Preston observed elsewhere in California, these first towns "were essentially local responses to locally perceived needs and opportunities, generated in close association with productive hinterlands."[21] When Indian conflicts

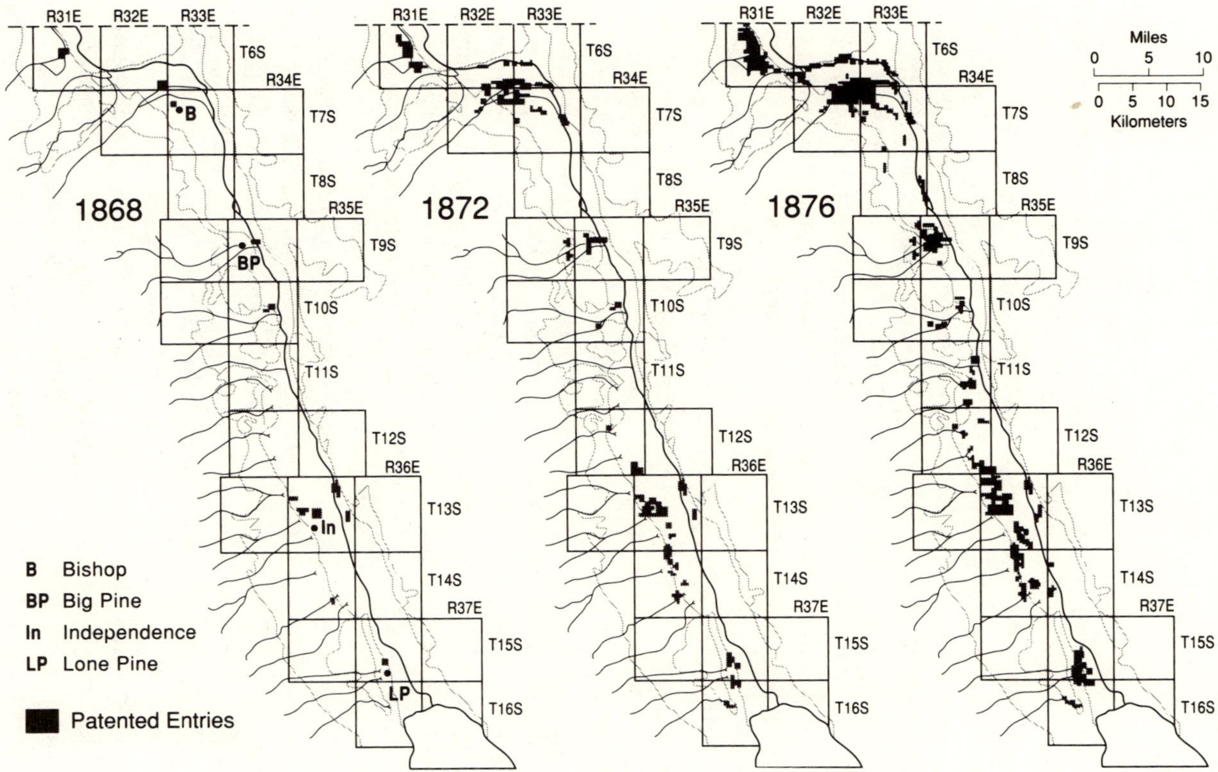

Figure 5.3. Patented entries in the Owens Valley, 1868–1876. (Compiled from Independence District Land Office tract books; redrawn from Sauder, "Sod Land Versus Sagebrush," 410)

diminished and the importance of locating near Camp Independence declined, a new focus of development began to emerge in the northern end of the valley. Between 1868 and 1872 the Bishop–Round Valley region received the majority of new settlers, and by 1876 this area had achieved dominance over all other foci of settlement.

Within this larger pattern, the sequence of maps of patented entries through 1876 shown in Figure 5.3 reflects human occupance based on the perceived capability of the land. The volcanic fields and other outcroppings that interrupt the flatness of the valley floor were obviously avoided, but also apparent is the preference of early settlers, as with the Paiute who preceded them, for locations on the Sierra side of the valley, with its perennial creeks cascading off the mountain slopes. Certainly the foremost concern of early settlers in this arid region was the ease with which water could be diverted to their newly planted fields. The former Paiute ditches assisted some in this task, but – borrowing from the custom developed by Western miners regarding the use of water on public lands – Owens Valley settlers also adopted the doctrine of appropriation. After determining a source for their irrigation water, pioneers posted and recorded notices identifying the location of stream diversion, the amount of water claimed, and the place and purpose of intended use. In accordance with the doctrine of appropriation, the first person to come to a stream and claim a portion of its flow had priority to exploit it, making it a form of personal property. Over time, a seniority system developed, the basic principal of which was "first in time, first in right." Unlike the riparian doctrine adopted by the more humid Eastern states, where water could legally be used only on land bordering a stream, prior appropriation rights allowed the appropriator to divert water in a stream to any location so long as it was put to beneficial use. By 1876 the largest of the valley's farming communities had developed in the Bishop–Round Valley region, primarily because more water was available for appropriation in this expanded end of the valley than elsewhere. The imprint of human activity in the West Bishop region highlights (Fig. 5.4) this historical relationship between lateral stream flow and settlement.

Yet by this time the continuing increase in the agricultural population of the Owens Valley began to be questioned by the *Inyo Independent,* which reported that "the cheaply available water for irrigating purposes is almost taken up in its use."[22] In early 1877, in attempting to promote the spread of larger irrigation canals in the valley, the newspaper again emphasized that nearly "all the available water . . . afforded by the numerous lateral streams of pure mountain water, is now under

Figure 5.4. Settlement in the West Bishop region is etched into the landscape between the north and south forks of Bishop Creek. (U.S. Department of the Interior, Bureau of Land Management, 1977)

claim."[23] There was still an abundance of land suitable for irrigation in the valley, but its utilization was predicated on the diversion of water from the Owens River, which involved a more organized undertaking that would not begin in any significant way until the mid-1880s. Within fifteen years after the first permanent settlers had filed their claims in the Owens Valley, the initial phase of settlement was over, and the chronological mapping of patented entries clearly illustrates the close relationship between early land selections and the availability of water for irrigation. Adaptations to more subtle landscape features in this remote desert valley, however, are less apparent, and a closer examination of Von Schmidt's field notes, as well as the valley's land cover reveal other important aspects of the pioneers' perceptions of their new home.

The Federal Land Survey and Settlement

Since field notes and township plats of federal surveyors were placed on file in the district land offices in order to help guide newly arriving settlers with their land selections, these documents probably had some influence on the location of early land entries. In fact, there is a clear correlation between soils that Von Schmidt classified as first-rate and initial land selections (Fig. 3.1). The most extensive area of first-rate soils in the valley, according to Von Schmidt, was located south of Poverty Hills in T12S R34E and T13S R35E. Von Schmidt concluded that the latter township held the greatest promise. "The quality of land in this Township is considerably above the common average," he wrote. "Almost the entire Township is covered with fine grass, the soil nearly all 1st Rate."[24] This is the township in which Camp Independence was located, and both the fort and the perceived fertility of the soil probably influenced the location of the earliest preemption and homestead claims in the valley. Yet Von Schmidt erred in his judgment regarding T12S R34E just to the north. There he noted: "The land in this Township is mostly level. Soil on a general average, 1st Rate, with the exception of the Western tier of sections, which is 3rd Rate. Most all of this Fractional Township is covered with fine grass. It has also a number of fine streams of water, passing through it in an Easterly direction."[25] Most of the designated first-rate soil in this township, however, remained unoccupied in 1876 because its lush natural meadows were waterlogged. To reclaim them would have required considerable labor and expense. Although soil ratings were designed to reveal the agricultural potential of the land, the practicalities of pioneer farming were often either misunderstood or overlooked by the field surveyors.

Ultimately, the natural meadows distributed along the west side of the valley were occupied first by Owens Valley settlers. California's state geologist, Josiah Whitney, during his reconnaissance to inspect the damage resulting from the 1872 Owens Valley earthquake, noted that "the ranches and settlements in the valley are chiefly at the lower edge of the sage-brush slope, on the alluvial patches, where the soil is soft and springy, and permeated by moisture from the mountain streams."[26] These are grass-covered areas identified by Von Schmidt as also having first-rate soils and, by implication, the best agricultural potential (Fig. 3.1).

To the north, Von Schmidt made favorable references to abundantly watered, level sections of land in T6S R31E, T6S R32E, T6S R33E, T7S R32E, T7S R33E, and T9S R34E.[27] In this region he noted the fields of nut grass under irrigation by the Indians. Regardless of his interpretation of soil quality in the northern half of the valley compared to the southern, early settlers soon recognized that the north end held the greatest promise. To the pioneer settler this area's greater width, abundant water supply, and, most important, the fields of nut-grass formerly irrigated by the Owens Valley Paiute meant that the land was fertile and that water could be diverted here with little effort. Von Schmidt's numerous references to Paiute irrigation in these northern townships no doubt induced early settlers to take a closer look at this end of the valley.

Sod Land Versus Sagebrush

Aside from information contained in the field notes and township plats of the federal surveyor, there was little to guide the pioneer settler, who was generally unfamiliar with farming in an arid environment. Obviously, water availability was a critical factor in any land selection, but soil fertility would also have been an important consideration. The former could easily be discerned through on-site inspection, but subtle differences in soil quality were more difficult to ascertain. Von Schmidt's classification of soils was crude at best, especially considering the diversity of soils actually found in the region. Lacking a detailed soil survey to assist them with their decision regarding where to locate, the first wave of settlers had to rely on their intuition, experience, and ability to adjust their land claims to fit the desired surface characteristics. Thus, an examination of the relationship among surface features, particularly the variability in land cover, will shed additional light on the manner in which early settlers evaluated the natural conditions of the Owens Valley.

In the Bishop–Round Valley region—the most intensively settled area of the Owens Valley during the early pioneer period—settlers encountered a variety of soils, but the predominant natural vegetation that the soils supported can be broadly classified as grass and sagebrush.

Table 5.1. The Relationship of Patented Acreage to Land Cover in the Bishop and Round Valley Regions Through 1876

	Through 1872		1873–74		1875–76	
	Round Valley	Bishop Region	Round Valley	Bishop Region	Round Valley	Bishop Region
Patented acreage	2,270	7,080	1,395	3,185	1,435	2,145
Percent grass-covered	80	82	52	81	20	61
Percent sagebrush-covered	20	18	48	19	80	39

SOURCES: Compiled from Von Schmidt, Field Notes, Watson and Storie, *Soil Survey of the Bishop Area, California,* and Independence District Land Office tract books.

The grass-covered fine sandy loams found on the lower margins of alluvial fans where perennial streams spread out over the valley floor were initially sought after. Fully 80 percent of the patented acreage through 1872 in Round Valley, and 82 percent in the Bishop region, consisted of grassland habitats (Table 5.1 and Fig. 5.5). These areas likely reminded early settlers of the grasslands of the western prairie states from which many had migrated and which others had crossed during their trek westward. Although not regarded as the best soils of the valley today, in the eyes of the pioneer settler they appeared to hold the greatest agricultural potential.

Most early land selections were 160 acres in size, the maximum unit available to settlers. The typical quarter-section could take on a variety of shapes, but the simple, even-sided square was the most popular, particularly on the flat valley floor, where relatively large blocks of sod land were found (Fig. 5.5). Where favored soils were not so expansive, parcels shaped like an **I**, **L**, **T** or **Z** were selected in order to encompass as much desirable land as possible. The shape of the pioneers' land selections is therefore linked to their appraisal of land quality. Less than 20 percent of the patented acreage in the Bishop–Round Valley region in 1872 encompassed land that supported sagebrush, and most of the sagebrush-covered acreage was simply unavoidably enclosed in land selections located on the margins of the grasslands (Fig. 5.5). Based on the earliest land locations in the northern end of the valley, it is evident that settlers' views of soil capability were influenced by the vegetation cover and that sagebrush was initially taken to indicate lower potential productivity when compared with the Owens Valley's sod lands. This initial belief about land quality was reinforced in an early issue of the *Inyo Independent*, which stated that, "with the exception of a considerable strip of land along Owen's river, and some fertile spots at the entrance of the ravines that make up into the Sierras, the country is . . . nothing but a sage barren, producing, besides the *artemesia* [*sic*], only a little bunch grass, with a few shrubs almost as worthless as the sage itself."[28]

Because the moist grasslands of the Bishop–Round Valley region were quickly taken up, later arrivals had no choice but to look to the sagebrush-covered slopes for their livelihood. This process began in Round Valley during the land selections of 1873 and 1874, when the sagebrush-covered acreage selected by new arrivals amounted to 48 percent of the total, while the more extensive meadows around Bishop continued to be taken up (Table 5.1 and Fig. 5.5). In the following two years, 80 percent of the acreage selected in Round Valley, and 39 percent of that located near Bishop, encompassed sagebrush-covered soils. The following account describes the experience of one pioneer who was forced to locate higher on the sagebrush slope:

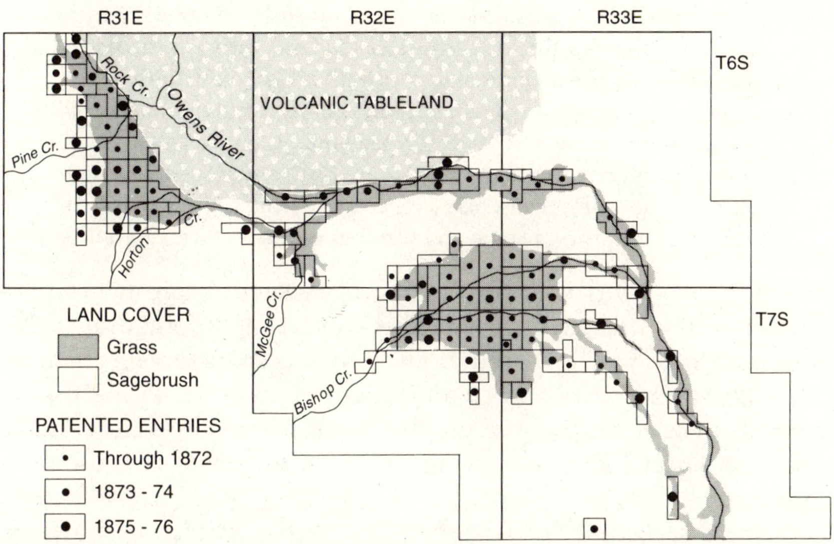

Figure 5.5. The relationship between patented entries and land cover in the Bishop–Round Valley region through 1876. (Compiled from Von Schmidt, Field Notes; Watson and Storie, *Soil Survey of the Bishop Area;* and Independence District Land Office tract books; redrawn from Sauder, "Sod Land Versus Sagebrush," 413)

> He was late in the valley. . . . [T]he level sections in the valley had long since been taken up. . . . He selected a quarter section of sandy looking ground lying . . . higher than the farmers farther down the valley. . . . [I]t suited [him] fine, all except the desolate appearance of his 160 acres, the sickly green sagebrush, and the numerous boulders scattered in desultory monotony over his land, which seemed most unpromising.
>
> Besides meagre household effects, he purchased a team and necessary farming machinery . . . and by spring he had much of it cleared; the sagebrush burned, the boulders neatly stacked up. . . . He had secured water rights in a creek that passed some distance above, and constructed a canal to carry water to his land.[29]

The reluctant settlers of sagebrush-covered land, however, eventually proved it was worthy of a closer look. Once their productivity was recognized, the "sage barrens" of previous years gradually gave way to a more favorable view. By 1877 this altered evaluation of the valley's sagebrush habitats was being promoted by the *Inyo Independent*:

> There is a vast amount of Government land yet unoccupied in this valley, eminently well suited for pasturage, and for fruit, cerials [*sic*], in short for every agricultural production. . . . In the main it consists of what is known as black sage brush land, a rich, sandy loam, which, when cleared of sage brush can be broken up with the ordinary team, in this differing from the heavy grass or sod land, hitherto the first choice, which requires a plow and team at first like those for the western prairies. The sage brush land, or much of it, is fully as fertile and no less desirable than the best of the sod land; it, however, requires more water, and is perhaps less durable. Being the freest in alkali, and variations in its component parts, its crops, when properly irrigated, are the most uniform in stand and turn out one acre with another.[30]

For potential settlers who were unfamiliar with this unusual form of vegetation, the *Independent* offered the following information: "this sage brush (*artemesia*) [*sic*] is a shrub rarely exceeding four feet in hight [*sic*], wide spreading branches, stout, gnarly trunk, and so superficially rooted as to be easily grubbed up or torn out with a horse and suitable appliance; burns like pitch pine in a stove or furnace, and is useful in ditch fencing."[31]

Not all sagebrush lands earned a favorable evaluation, however, even after their overall productivity was recognized. Settlers showed some selectivity in their assessments, entering only areas that supported a healthy stand of sagebrush and avoiding areas with a scanty growth.[32] The hardpan soils located between Bishop and Round Valley, for example, supported only a sparse cover and were bypassed by pioneer settlers because they were accurately perceived as having little agricultural potential. Nevertheless, extensive sagebrush habitats in the region were eventually brought under the plow, and because they were free of alkali and could easily be irrigated, some sagebrush-covered soils later came to be regarded as among the most desirable in the valley.[33]

By the end of the 1870s an empirical testing of the quality of the land, farm by farm, in the Owens Valley had been underway for a decade and a half, and the location of fertile and marginal lands had largely been defined. Undoubtedly influenced by early accounts of the region, pioneer settlers immediately took up the moist sod lands of the valley. Having originally come from more humid areas, the valley's lush meadows must have appeared to early settlers as green oases among the desolate "sage barrens." The pioneers therefore initially attempted to encompass as much sod land as possible within their quarter-section claims. Certainly a more favorable evaluation of sagebrush habitats would have been reflected in a higher proportion of patented sagebrush-covered acreage during the early years of settlement. In time, however, the valley's farming frontier expanded out to the leading edges of the sagebrush habitats, and pioneers who arrived later in this isolated region where the wilderness had already been partially transformed were more willing to experiment. Only then did it become evident that the sagebrush-covered lands were just as productive and more easily worked than the valley's sod lands. Commonly, sagebrush land adjacent to meadowland was entered first, and then the farming frontier expanded outward in a contiguous wave into the sagebrush habitat. In the Owens Valley, as elsewhere, the process of experimentation was a major part of the frontier experience, but it began to unfold only after the more familiar habitats were largely occupied.

6. Establishing a Working Economy, 1860–1880

> Even our own citizens who have had no recent opportunities for direct observation, would feel astonished were they to look at our farms, particularly through Bishop and Round Valley.
>
> — "Among the Farms," *Inyo Independent*, July 28, 1877

A Midwestern Farm Complex

Rural patterns of livelihood that initially evolved in the Owens Valley were influenced by economic activities that prevailed in the western Great Basin during the 1860s and 1870s, by the valley's isolation from the principal nuclei of development in the Far West, by the source regions of early settlers, and by differences in environmental conditions from north to south in the valley. For those who first settled in the Owens Valley, and for others who followed, the problem was how to transform this isolated region into a working economy. Essentially, the economy of the frontier community that evolved within this narrow trough relied on supplying provisions and services to Cerro Gordo and other mining locations in the western Great Basin, and the initial impact of these mining centers on agricultural development in the Owens Valley was to create a demand that far exceeded the supply. This was the stimulus that encouraged farmers to overcome the harsh arid geography of the region and inspired them to carve out permanent communities on the valley floor. In contrast to its previous transient nature, an air of stability enveloped the valley, and although the hard times that followed each mining boom would reduce the agricultural interests, they would never entirely eliminate them.

The federal census of 1870 provides an official record of the progress of farm development in the Owens Valley within a few years after permanent settlement began.[1] The majority of the valley's farm population in 1870 derived from the northeastern quadrant of the country (Table 6.1). At that time the place of origin of 63 percent of the farmers was a Midwestern, Middle Atlantic, or New England state, with the largest number coming from Ohio. By 1880 the valley's farm population had more than tripled, and the source areas were somewhat clearer. The Midwestern states of Ohio, Illinois, and Missouri were important source regions, as were New York and Pennsylvania in the Middle Atlantic region, Maine and Vermont in New England, and the British Isles and Germany among foreign countries. It is evident that many of the settlers from the Middle Atlantic region and New England, as well as many foreign-born settlers, had spent some time in the Midwest before embarking for California, a fact confirmed in the census by the place of birth of the children of many Owens Valley pioneers.

Since the majority of Owens Valley settlers had lived in the Midwest at one time or another, it is no coincidence that the small farm, mixed crop and livestock farming complex – the hallmark of Midwestern farming – was initially transplanted to the valley. This system, whose original source was southeastern Pennsylvania, had become well-established in southwestern Ohio by 1820.[2] It depended on a standard three-crop rotation of corn (the principal crop), small grains (generally wheat or oats), and hay for fattening livestock. This farming complex could be reproduced in the arid West because irrigation offset rainfall deficiencies. Corn thrived on the valley's moist sod lands, and in 1870 it was by far the leading grain crop produced in each of the Owens Valley's subregions, while wheat was the primary companion grain (Table 6.2). Alfalfa hay was also planted and, along with natural meadow grasses, cut and stored for winter feed (Table 6.3). Nearly all the farmers had some horses, milk cows, range cattle, and hogs. Horses were more numerous in the Bishop–Round Valley area, reflecting the importance of teaming to the Mono County mines. Most of the range cattle grazed on the salt grass meadows of the public domain in the southern half of the valley. Sheep raising had not yet been widely adopted by Owens Valley farmers, although flocks of four to eight thousand head from the San Joaquin Valley regularly embarked each spring on the long trail through the Owens Valley to higher summer pasture in the Sierra, returning in the fall. At Owens Lake, where the trail narrowed, an important source of county revenue was derived from the fee levied on all bands of passing sheep.[3]

As in the Corn Belt, corn and hay were fed to livestock, while wheat was the principal cash crop. At this time, new mining centers were

Table 6.1. Farm Population in the Owens Valley by Place of Origin, 1870 and 1880

Place of Origin	Farm Population 1870	Farm Population 1880
MIDDLE WESTERN STATES		
Ohio	14	25
Illinois	6	15
Missouri	3	10
Indiana	0	6
Michigan	2	6
Iowa	0	4
Wisconsin	0	1
Total	25	67
MIDDLE ATLANTIC STATES		
New York	9	26
Pennsylvania	4	11
New Jersey	0	2
Delaware	0	2
Total	13	41
NEW ENGLAND STATES		
Maine	2	16
Vermont	5	10
Massachusetts	0	3
Connecticut	3	3
Rhode Island	1	2
Total	11	34
SOUTHERN STATES		
Virginia	2	8
Tennessee	3	7
Kentucky	2	6
Arkansas	1	4
North Carolina	0	3
Texas	0	3
West Virginia	0	2
South Carolina	0	1
Georgia	0	1
Alabama	0	1
Total	8	36
FAR WESTERN STATES		
California	0	8
Nevada	0	1
Total	0	9
FOREIGN COUNTRIES		
British Isles	7	24
Germany	4	20
Canada	1	7
Total	12	51
Unidentified	4	4

SOURCES: U.S., Ninth Census, 1870, Manuscript Census of Inyo County, Schedule 1: Inhabitants, and Schedule 3: Productions of Agriculture; U.S., Tenth Census, 1880, Manuscript Census of Inyo County, Schedule 1: Inhabitants, and Schedule 2: Productions of Agriculture.

NOTE: Place of origin from Schedule 1 was counted only for those individuals who appeared in Schedule 3 in 1870 and Schedule 2 in 1880.

Table 6.2. Grain Production by District in the Owens Valley, 1870 and 1880

	Corn (bu.)		Wheat (bu.)		Oats (bu.)	
District	1870	1880	1870	1880	1870	1880
Round Valley	4,645	5,663	3,571	12,520	1,145	16,600
Bishop	5,300	16,271	4,511	8,835	710	5,987
Big Pine/Fish Springs	5,940	3,111	1,034	1,276	70	–
Independence/ Lone Pine	7,030	6,066	4,513	6,202	250	–
Owens Valley Total	22,915	31,111	13,629	28,833	2,175	22,587

SOURCES: U.S., Ninth Census, 1870, Manuscript Census of Inyo County, Schedule 3: Productions of Agriculture; U.S., Tenth Census, 1880, Manuscript Census of Inyo County, Schedule 2: Productions of Agriculture.

Table 6.3. Hay and Livestock Production by District in the Owens Valley, 1870 and 1880

	Hay (tons)		Horses		Milk Cows	
District	1870	1880	1870	1880	1870	1880
Round Valley	172	1,101	88	565	122	164
Bishop	457	3,310	728	1,307	316	400
Big Pine/Fish Springs	50	1,256	91	451	60	155
Independence/Lone Pine	747	2,007	508	857	288	305
Owens Valley Total	1,426	7,674	1,415	3,180	786	1,024

SOURCES: U.S., Ninth Census, 1870, Manuscript Census of Inyo County, Schedule 3: Productions of Agriculture; U.S., Tenth Census, 1880, Manuscript Census of Inyo County, Schedule 2: Productions of Agriculture.

Barley (bu.)		Total (bu.)	
1870	1880	1870	1880
460	2,528	9,821	37,311
596	16,418	11,117	47,511
765	3,223	7,809	7,610
3,084	9,728	14,877	21,996
4,905	31,897	43,624	114,428

Range Cattle		Sheep		Swine	
1870	1880	1870	1880	1870	1880
531	380	–	3,500	188	631
664	1,956	–	15	237	893
191	1,635	–	4,207	71	257
3,342	1,498	521	2,000	170	527
4,728	5,469	521	9,722	666	2,308

Table 6.4. Miscellaneous Farm Products by District in the Owens Valley, 1870 and 1880

District	Potatoes (bu.) 1870	Potatoes (bu.) 1880	Orchard Products (dollars) 1870	Orchard Products (dollars) 1880	Market Garden Produce (dollars) 1870	Market Garden Produce (dollars) 1880
Round Valley	1,057	4,021	–	150	–	200
Bishop	2,724	9,447	–	700	–	2,007
Big Pine/Fish Springs	210	3,876	–	2,661	–	50
Independence/Lone Pine	2,302	5,509	300	1,000	1,550	2,500
Owens Valley Total	6,293	22,853	300	4,511	1,550	4,757

SOURCES: U.S., Ninth Census, 1870, Manuscript Census of Inyo County, Schedule 3: Productions of Agriculture; U.S. Tenth Census, 1880, Manuscript Census of Inyo County, Schedule 2: Productions of Agriculture.

continually opening in Inyo County, thereby expanding both the demand and the cash market for the valley's produce. Wacouba, Darwin, Panamint, and Cerro Gordo had thriving silver, gold, lead, and zinc mines. Surplus produce from the valley could only partially fill the demand of these mining communities. Yet, while most of the surplus crop and livestock production of the valley found its way to the nearby mining districts, a local market developed concurrently in the vicinity of Bishop, Big Pine, Independence, and Lone Pine, the valley's emerging agricultural service communities. Potatoes, milk, and butter – produced on nearly every farm – were important early staples, and some orchard and market garden production became established in the Independence region, primarily to serve the soldiers of Camp Independence (Table 6.4). Provisions from the outside were difficult to secure in this isolated frontier, and residents therefore traded heavily among themselves.

Sodbreaking, planting and irrigating entailed hard work, and initially a quarter-section parcel was more than enough acreage to attempt to bring under the plow. In 1870 the average farm size was 176 acres, but 89 percent of all farms were no more than 160 acres in size.[4] Only 39 percent of all farm acreage was classified as improved, and most of this was simply permanent meadows (Table 6.5). More farms, more total farm acreage, and more improved acreage were recorded in the

Butter (lbs.)		Honey (lbs.)		Poultry (no.)		Eggs (dozens)	
1870	1880	1870	1880	1870	1880	1870	1880
3,405	6,073	na	1,035	na	754	na	2,902
5,400	19,655	na	5,260	na	2,655	na	11,127
2,500	1,400	na	1,780	na	1,412	na	3,342
7,265	12,353	na	3,755	na	2,958	na	5,778
18,570	39,481	na	11,830	na	7,779	na	23,149

Independence–Lone Pine district than elsewhere in the valley,[5] thereby confirming that the southern portion of the Owens Valley initially developed more quickly than the north. The Bishop region ranked second in total number of farms and total farm acreage, followed by Round Valley and the Big Pine–Fish Springs regions.

Because demand exceeded supply and because prices for farm products in the mining communities were high, to maximize production most pioneer farmers used hired labor for clearing land, digging ditches, planting, cultivating, and harvesting. The Owens Valley Paiute, most of whom were desperate for some kind of livelihood, took up jobs on the farms and ranches. Paiute men supplied most of the seasonal field labor, while Paiute women worked as domestic servants, although they also specialized in winnowing the harvest.[6] In 1870, 1,150 Indians were counted in the valley. Although their tribal ways and subsistence patterns had been severely disrupted, they were able to restore their family patterns, and most lived in traditional shelters clustered into rancherias.[7] The camps were generally located either on the margins of farms or at the edge of the four principal towns, but the largest was at Camp Independence.[8] At each of these sites the Paiute were granted land-use rights for small garden plots. Paternalism defined the relationship between the Paiutes and their employers. According to John Walton,

Table 6.5. Number of Farms, Farm Acreage, and Improved Acreage by District in the Owens Valley, 1870 and 1880

	Number of Farms		Average Farm Size (acres)		Acreage in Farms		Improved[a] (Irrigated) Acreage		Improved Acreage as % of Total Farm Acreage	
District	1870	1880	1870	1880	1870	1880	1870	1880	1870	1880
Round Valley	12	29	160	199	1,920	5,767	668	4,021	35	70
Bishop	24	108	187	184	4,500	19,870	1,268	11,045	28	56
Big Pine/Fish Springs	5	34	160	188	800	6,380	395	3,552	49	56
Independence/Lone Pine	32	51	177	273	5,669	13,915	2,657	7,643	47	55
Owens Valley Total/Average	73	222	176	207	12,889	45,932	4,988	26,261	39	57

SOURCES: U.S., Ninth Census, 1870, Manuscript Census of Inyo County, Schedule 3: Productions of Agriculture; U.S., Tenth Census, 1880, Manuscript Census of Inyo County, Schedule 2: Productions of Agriculture.

[a]Improved acreage, according to the census, included permanent meadows and pasturelands.

"Indians were adopted as childlike servants and given a name like any new addition to the family."[9] Consequently, the Paiutes commonly accepted the family name of their employer.

In the early 1870s, discussion surfaced at the federal level about once again removing the Indians from the valley, this time to a proposed reservation on the opposite side of the Sierra along the southern Tule River.[10] In contrast to previous years, the settlers now objected to Indian relocation because it meant the loss of their indigenous labor force. The *Inyo Independent*, reflecting the opinion of the majority of the valley's farmers, argued against the proposed government action:

> Their [Indian] labor is of essential importance in public economy, and supplies a want in that direction that no other can do as well. On this account their presence is a general benefit and the people here will universally protest against their removal. . . . These Indians as a tribe are industrious and contented, peaceable and well disposed. . . . All that citizens here ask is that the Department will see that these Indians are not interfered with at all by reservation agents in any manner. . . . The value of, and demand for their labor, both field and household, insures them good treatment at the hands of the whites, and a support free of cost to the Government.[11]

A cheap labor force was now an integral part of the pioneer economy, which Walton describes as being "built on the labor of whites and Indians working together in an exploitative yet closely interdependent social relationship."[12] Indeed, the Paiute had adjusted to farm life so readily that in the mid-1870s some began to file on homesteads of their own under the Indian Homestead Act of March 3, 1875.[13] The relocation idea was subsequently dropped.

In 1865, Josiah Earl preempted 160 acres near Camp Independence, and his farm exemplifies the scale of operation and variety of farm output among the successful early pioneers in the Owens Valley (see Table 6.6). The following account provides a glimpse of pioneer life on the Earl farm, as described by Josiah Earl's son in *The Enchanted Valley*:

> [My father] filed on a tract of government land next to Fort Independence . . . where a hundred soldiers were stationed to protect against the Indians, recently hostile. He started to build quite an ambitious adobe house on the land, and when the lower walls were up, a board lean-to was erected against the rear, into which we moved from our "palatial" adobe in the sagebrush. . . .
>
> In the spring, after we moved into the lean-to, my father . . . [began] to plow. In a few months, our vegetables matured and we harvested our

Table 6.6. Josiah Earl's Farm, 1870

ACRES	
Improved	80
Unimproved	80
Total in farm	160
FARM VALUES AND WAGES PAID	
Value of farm	$2,500
Estimated value of all farm production	$4,250
Total amount of wages paid, including value of board	$1,320
LIVESTOCK	
Horses	6
Milk cows	10
Range cattle	10
Sheep	20
Value	$1,100
FARM PRODUCTS	
Corn	1,071 bushels
Barley	208 bushels
Potatoes	223 bushels
Value of market garden produce	$750
Butter	365 pounds

SOURCE: U.S., Ninth Census, 1870, Manuscript Census of Inyo County, Schedule 3: Productions of Agriculture.

wheat, corn and potatoes and had an abundance of flour and meal therefrom.

Within a few steps of the kitchen door of our home was the cellar, excavated to a depth of about six feet, with the timbers upholding the gabled roof extending about two feet above the surface. The structure was about twelve by twenty, with stone steps descending to its entrance. In the cellar, cooled by water running through it in a flume, we kept the milk, flour, bread, meat and such clean products, but had a different one of somewhat similar design where we kept potatoes and other coarser supplies.

We [also] raised . . . all kinds of melons . . . cabbages, onions, lettuce, cauliflower, in fact, all kinds of vegetables. Finally we raised a good deal of fruit and berries.

Early in our stay in the Valley an itinerant brought in a wagonload of active beehives. We bought one, and by the time I left the Valley [in 1872] we were maintaining thirty-six hives, all descendents of the first. Our honey was in great demand and much of it was of such quality that I have never seen its equal. . . . [It had] a most exquisite flavor from wild buckwheat out of which it was made.

We had a ready market for all the truck garden stuff and other produce of the farm, because the miners at the camps were glad to get it . . . [and] we sold a good deal of butter and large numbers of eggs and chickens. This . . . was . . . clear gain, for the cows grazed on the public lands and got fodder from the field after the corn ears were removed and, for a short time in the winter, were fed some alfalfa hay from the farm. We kept about four hundred chickens and fed them corn and wheat from the ranch and the skim-milk, butter milk, and they foraged a good deal. The horses were supported by hay and grain from the farm, and we took care of them, so they cost practically nothing. . . .

We also sold at big prices hogs, calves and beef animals.[14]

Farming in a Grazing Frontier

As land continued to be brought under the plow during the early 1870s, and as new claims were filed on the public lands that stockmen had formerly used as open range, the destruction of growing crops by free-ranging stock increased. The transition to farming in this grazing frontier did not go smoothly, as conflicts inevitably developed between the valley's farmers and stockmen. Paradoxically, pioneer farmers of the valley suffered a fate similar to that which they had inflicted on the Paiute in the previous decade. The difficulties associated with this transition are revealed in the following experience of two pioneers who attempted to farm the lush meadowland of Round Valley:

> John E. and Thomas E. Jones decided to undertake ranching in Round Valley, and sowed the first wheat there in the spring of 1865. By borrowing from others, enough of a team was made up to laboriously break up 4½ acres of heavy sod. In this soil . . . wheat . . . was sowed. Five thousand head of livestock were being pastured in the valley, and in mid-June the herds invaded the little ranch and destroyed the crop. Jones replanted his tract, made a shelter of willow boughs and a wagon sheet and alone watched day and night until his wheat ripened.[15]

Fencing would be needed to keep stock out of the newly sown fields, but in an arid, treeless region such as the Owens Valley, fence materials were difficult and expensive to acquire. "It is quite beyond the means of the majority of our farmers to put secure fences around their places," reported the *Inyo Independent* in 1871, "yet the presence of a small band of stock in a neighborhood may cause a heavy expense."[16] When the Owens Valley began to be settled, barbed wire had not yet been invented, and when it did appear on the market in 1874, it cost twenty-five cents a pound, which was prohibitively expensive for most pioneer farmers.[17] Lacking other means of protecting their fields from roving bands of stock, many early farmers in the southern half of the valley resorted to "ditch" fencing, whereby sagebrush was piled up around a field and held down by dirt excavated along the outside perimeter.[18] If fence posts were available, they would be inserted at intervals in the brush and connected by a single rail. Farther north, in the Big Pine, Bishop, and Round Valley districts, some farmers attempted to enclose their fields with sod fences.[19]

To increase the supply of fencing material, black locust trees were introduced into the valley in the late 1860s,[20] and cottonwood trees were planted for the same purpose.[21] But the establishment of tree plantations for fencing material gained little acceptance in the valley. More commonly, locust and cottonwood trees, as well as lombardy poplars, were planted as windbreaks around fields to protect newly planted crops from harsh spring winds. There was little time for extensive tree culture and fence building when most pioneer efforts were devoted to breaking and planting new ground. Hence, in 1872 the *Inyo Independent* noted that there was "scarcely a securely fenced farm in the county, and, in most cases, it is hardly possible to make them so."[22]

The farmers in the southern region of the Owens Valley were most severely affected by free-ranging stock, and they soon began to push for legislative approval of a "no-fence" law for Inyo County that would hold stockmen liable for damage to unprotected fields. By the early 1870s, livestock from the Bishop–Round Valley region was being herded north into Long Valley for summer grazing, largely in response to the approval of a local no-fence law protecting northern valley farmers.[23] In calling for legislative approval of a no-fence law encompassing the entire county, the *Independent* argued that without such a law "the owner of a few head of stock may, and often does, virtually monopolize lands sufficient for homes for dozens of poor families . . . [that] will continue to lie in idleness for years to come . . . and eventually be gobbled up in large tracts by speculators."[24] Farmers of the region believed that fence building should be undertaken to keep their own stock in but not to keep those of others out. Stockmen countered by suggesting that a no-fence law, if enacted, would enable a single farmer to cultivate a few acres in various places, thereby excluding stock from an extensive range.

In the spring of 1872 the state legislature ratified a compromise no-fence measure that affected Inyo and several other California counties. It required stock raisers to use mountain meadows instead of valley lands for summer pasture for seven months during the growing season, but it allowed free range for the remaining five winter months when there were no crops exposed.[25] The new provision enabled farmers to confiscate stray cattle and to sue ranchers for damage to their crops during the summer regardless of whether their fields were fenced. But winter stubble fields continued to be common property under this measure unless protected by fences. Farmers in the Owens Valley and elsewhere in California quickly recognized that free-ranging stock in the winter could inflict nearly as much damage to fields and irrigation systems as before.[26] Consequently, in 1875 the so-called Fresno no-fence law was substituted for the original measure. Now stockmen were held responsible for the destruction of farm property, fenced or unfenced, throughout the year.

Achieving Agricultural Dominance in the North

The approval of a no-fence law for Inyo County was symbolic of the importance that farming had achieved in the Owens Valley by the mid-1870s. The Bishop–Round Valley region had by now attained agricultural dominance over the other districts. In March 1876 it was reported that Bishop "is the champion agricultural district in our county."[27] The following year the *Inyo Independent* provided a contemporary description of agricultural accomplishments in the northern end of the valley:

> In [Round Valley] . . . where small grain is more of a specialty, there are numbers of fields of wheat, barley and oats, containing from fifty to one hundred and fifty acres, where the grain, yet of a deep green, stands as thickly on the ground and as tall, even, and well headed as in the most favored grain growing localities of the coast counties.
>
> Bishop Creek valley proper is far more diversified in soil and productions, and covers a much greater area. Bishop is more given to corn than wheat, but seems to be quite as well adapted to one as the other.
>
> Down nearer home is Big Pine . . . [where] a field of [alfalfa] . . . would command admiration, even from a Kern Islander. So would certain corn fields.[28]

The newspaper also offered its conclusions about the principal reason for the more impressive improvements in the north: "Notwithstanding the dryness of the season, and the short supply of water for irrigating purposes in the valley south of Big Pine, at that place and at Bishop and Round Valley all requirements are satisfied, and in each there is yet water running to waste, enough, it would appear, to irrigate double the area now under cultivation."[29]

The federal census of 1880 confirmed the above developments. Overall, the number of farms in the Owens Valley had more than tripled during the decade of the 1870s, while Bishop had overtaken all other communities in nearly every farm-related category (see Tables 6.2–6.4). By 1880 it was unquestionably the premier agricultural district in the valley, with more than double the number of farms of any other region and significantly more total farm and improved farm acreage (Table 6.5). Throughout the valley, however, the quantity of improved acreage increased during the 1870s as the ratio of improved to total farm acreage rose from 39 to 57 percent. Still, only a small proportion of this improved acreage was actually tilled; for the purposes of the census, permanent meadows and pastures were considered "improved" or "agricultural" land. The Inyo County assessor's report for the 1877 fiscal year, for example, counted only 4,245 cultivated acres in the valley, excluding hay production.[30] A trend toward farm enlargement was also underway, particularly in the valley's southern end (Table 6.5). Despite this trend, in 1880, 70 percent of all farmers in the Owens Valley were operating farms of 160 acres or less in size.[31]

Modifying the Farm System

Mixed farming was still the norm in 1880, but the prevailing economic forces in the region, and its arid geography, prompted farmers to modify the original system. Now barley, in great demand in nearby mining centers, narrowly outranked corn as the leading grain crop raised in the Owens Valley, while wheat took third place (Table 6.2). Adaptations to local environmental conditions resulted in some regional specializations as well. In the higher elevation of Round Valley, oats and wheat were the leading cereal crops, whereas barley and corn were the principal products of the Bishop region. No oat fields were found from Big Pine south to Owens Lake, whereas barley remained the dominant grain crop of the southern end of the valley.

As in the past, alfalfa hay was grown in every section of the valley, and its yield increased significantly during the 1870s (Table 6.3). Because of its ease of production, its superior quality, and the expanding demand for hay by livestock interests in the valley, alfalfa was taken out of rotation with grain crops and instead was grown on the same plots of ground each year, giving two to three cuttings per year. A two-crop rotation of corn and small grains therefore became the norm. Nearly all farmers still owned a few milk and beef cattle, as well as some horses and hogs, and several large cattle herds had been assembled by a few of the valley's ranchers. Sheep raising gained in prominence during the 1870s, with sizable bands found in Round Valley and at Big Pine–Fish Springs and Lone Pine (Table 6.3). In addition, butter, eggs, poultry, potatoes, orchard and market garden produce, and honey were important products of the valley. The Bishop region out-produced all others in nearly every category (Table 6.4), but Big Pine was clearly the valley's leader in fruit production, largely because the Owens Valley's pioneer orchardist had settled there.[32]

A Mining Bust in an Isolated Country

The impressive developments reflected in the 1880 census figures actually camouflaged the less than flourishing times that now descended over the valley. The year 1877 marked the beginning of many lean years in the Owens Valley. With a dramatic decline in the price of silver, the Cerro Gordo mines went into a decline, as did the other smaller mining camps that were accessible by teaming.[33] In addition, Camp Independence, consistently a convenient local market for surplus farm products, was abandoned in July 1877.

As discussed earlier, the valley's isolation from the more settled regions of California meant that it was almost totally reliant on the nearby markets for its prosperity. Although cattle, horses, sheep, and hogs could, if necessary, be driven out to distant markets, hauling such products as wheat, barley, corn, potatoes, butter, and eggs by wagon over long distances was simply not practical. The decline of Cerro Gordo and the closing of the fort brought an end to the Owens Valley's first period of prosperity.[34] In May 1877 the *Inyo Independent* reported that agriculture had "reached its maximum – seen its best days till a railroad opens a new market. . . . Every additional acre brought in cultivation becomes a detriment to the farmers when their previous productions were fairly equal to demands."[35] Many pioneers now began to look for a new life elsewhere.

Yet, surprisingly, new land continued to be brought under the plow in the northern end of the valley, and in 1880, with a hint of jealousy, the *Inyo Independent,* published in Independence, printed the following report:

> Whatever business depression the mining communities of Inyo County are suffering from, the fact does not appear to affect in the least the general prosperity of the farming country to the north of us. The productive lands, with abundance of water for irrigation, make Big Pine, Bishop and Round Valley veritable oases in the vast desert of mining districts which surround. During the present year many new farms have been placed under cultivation and much new land belonging to the older places broken up.
>
> The central portion of Bishop Creek exhibits many solid improvements in fencing and buildings. . . . [T]he valley to the east has so filled up with farms that the town stands nearer the center of population than before.[36]

With its larger population, the northern end of the Owens Valley was blessed with a more substantial home market and was therefore buffered to some extent against the consequences of the region's slumping mining activity.

The Plight of Three Farmers

A comparison of three pioneer farms, one located in Round Valley, another near Bishop, and the third in the Independence region, sheds additional light on the progress and problems associated with farming in the Owens Valley during the late 1870s (see Table 6.7). William Horton, using the right of preemption, squatted on a quarter-section of land in Round Valley in 1864 and purchased his farm from the government in 1873 with agricultural college scrip. As one of the earliest settlers in the region, Horton is credited with growing the first crop of corn and vegetables in Round Valley.[37] By 1870 the value of his farm was estimated to be $2,000. His livestock consisted of a few horses, a single milk cow, and a hog. His leading cash crop in 1870 was corn, but he also raised some wheat and oats and several bushels of potatoes. His farm production grossed $1,881 that year. By 1880 Horton had enlarged his farm to 245 acres, and its value had increased to $9,000. The number of livestock on his farm was greater than in 1870, but his income was still largely dependent on grain production. Experimentation over the years had demonstrated that oats did well at higher elevations, making it his leading grain crop, but corn also was an important crop. His farm products in 1880 brought in $2,590, but after factoring in $900 in wages plus his other expenses, his net income in 1880 was actually less than it had been in 1870.

With a somewhat different orientation in farm production, Thomas Hutchinson experienced a similar fate during the 1870s. Like Horton, he first squatted on a quarter-section parcel – on the sod land adjacent to Bishop Creek west of the present site of Bishop – in the mid-1860s and subsequently purchased his land with military scrip in 1873. By 1870 his farm was valued at $4,000. He was engaged mainly in raising horses for the teamsters serving the northern mining communities, but he also had relatively large herds of milk and beef cattle. The combined value of his stock was more than $14,000. Among his cash crops, corn was by far the most important. He also raised significant quantities of potatoes. His farm income in 1870 amounted to more than $3,000. Unlike Horton, Hutchinson did not enlarge his farm during the 1870s; instead, by 1880 nearly all of his acreage had been improved. This factor was no doubt responsible for Hutchinson's citation by the *Inyo Independent* three years earlier as having one of the finest farms in the Bishop region.[38] The value of his farm had nearly doubled over the decade. Livestock production was still important to him, but his emphasis had shifted more toward the range cattle business. He supplemented his income with wheat, corn, oats, and barley production, and sold potatoes and butter. The total value of his farm output in 1880 was about 15 percent above his 1870 farm income, but after subtracting wages and other expenses, he, like Horton, was little better off in 1880 than in 1870.

Jacob Vagt was also an early resident of the valley, having preempted a quarter-section claim near Camp Independence in 1864. He was a model pioneer farmer and was one of the first to produce what the *Inyo Independent* called "a most luxuriant growth of alfalfa . . . from low ground which, previous to plowing, was literally white with alkali efflorescence."[39] Vagt's farm embodied the mix of activities that symbolized pioneer farming in the 1860s and 1870s. He owned a few horses, milk cows, range cattle, and swine, and he raised wheat, barley, and some corn. In 1870 Vagt was one of the most successful farmers in the valley. He grossed more than $4,400 in that year, and even after deducting $1,500 in wages, his net income was higher than Horton's or Hutchinson's (assuming his other expenses were approximately equal to theirs). But by 1880 his farm operation had taken a significant turn for the worse. His livestock herds had been reduced in number and were now valued at only $250, and the quantity of improved acreage on his

Table 6.7. Sample Farms in the Round Valley, Bishop, and Independence Districts, 1870 and 1880

	William Horton (Round Valley)		Thomas Hutchinson (Bishop)		Jacob Vagt (Independence)	
	1870	1880	1870	1880	1870	1880
ACRES OF LAND						
Improved acres	60	165	75	150	120	90
Unimproved acres	100	80	85	10	40	70
Total farm acres	160	245	160	160	160	160
FARM VALUES AND WAGES PAID						
Value of farm	$2,000	$9,000	$4,000	$7,000	$1,500	$2,000
Estimated value of all farm production	$1,881	$2,590	$3,040	$3,500	$4,418	$1,300
Total amount of wages paid, including value of board	$50	$900	$600	$1,200	$1,560	$100
LIVESTOCK						
Horses	4	17	175	30	19	10
Milk cows	1	4	50	2	8	3
Range cattle	–	12	100	220	21	5
Swine	1	40	10	40	36	30
Total value of stock	$575	$800	$14,174	$3,000	$1,690	$250

FARM PRODUCTS						
Hay (tons)	18	50	23	30	50	90
Wheat (bushels)	65	233	50	450	250	306
Corn (bushels)	575	714	700	400	60	210
Oats (bushels)	150	817	100	400	–	–
Barley (bushels)	–	–	–	1,000	417	50
Potatoes (bushels)	160	100	150	266	100	34
Butter (pounds)	–	450	–	200	100	360

SOURCES: U.S., Ninth Census, 1870, Manuscript Census of Inyo County, Schedule 3: Productions of Agriculture; U.S., Tenth Census, 1880, Manuscript Census of Inyo County, Schedule 2: Productions of Agriculture.

farm had declined from 120 to 90 acres. His overall production of grain crops had also decreased, led by a drastic reduction in barley. The closing of Camp Independence no doubt impacted the income he received from his potato crop, although here, as elsewhere in the valley, butter continued to be produced in large quantities. The wages he paid to farmhands in 1880 came to only $100, yet he was still left with just $1,200 from which to deduct the remainder of his expenses. This compared to his net income of $2,900 in 1870.

The three farm operations illustrate that, although the north end of the Owens Valley was no longer booming, it continued to hold its own in the face of shrinking markets. On the other hand, the southern region, judging by Jacob Vagt's misfortune, was mired in a severe depression. Indeed, the southern districts would languish from this time forward, never reclaiming the prosperity of the late 1860s and early 1870s. The valley's farmers had overcome numerous obstacles to development by the early 1870s, including the suppression of the native population, the creation of a diversified irrigated farming system, and the resolution of conflicts between farmers and stockmen. But the most perplexing obstacle to the valley's sustained economic development – accessibility to the expanding West Coast markets – remained unresolved. The downturn in mining in the region in the late 1870s therefore left only a small local demand for the farmers' crops.

7. Economic Organization and Environmental Alteration, 1880–1905

It is now generally conceded that this is getting to be a hard country for a farmer to flourish in.

— "Letter from Round Valley" *Inyo Independent*, November 3, 1883

Very little government land now remains open for location, and none within lines of canals now being made. Those who want to find land they can have for nothing must search elsewhere.

— "Mistaken Land Hunters" *Inyo Independent*, December 3, 1887

The Carson and Colorado Railroad

The last two decades of the nineteenth century in the Owens Valley were characterized by a continuation of the economic conditions that had emerged in the late 1870s. With the collapse of the only market for the valley's surplus farm products, the Owens Valley languished in isolation, and its pioneer settlers began to leave.[1] The primary obstacle to recovery was the valley's inability to ship grain and other surplus products to distant markets. The unbroken granite wall of the Sierra Nevada was a formidable barrier, severing the Owens Valley from California's San Joaquin Valley and the Pacific Coast, for the mountains could not be easily crossed anywhere nearby. With the initiation of service by the Carson and Colorado Railroad in the summer of 1883, the valley was linked to the outside world by rail for the first time. But the new line met with little fanfare, as it did little to alleviate the valley's isolation from the more settled regions of California.

The extension of the Carson and Colorado into the Owens Valley was inspired by the region's mining activity, but by the time it reached the Owens Valley, the initial period of mining excitement was largely over. The railroad was the brainchild of the builders of the Comstock bonanza road, the Virginia and Truckee Railroad. Diminishing revenues from the operation of the Virginia and Truckee caused the owners to look toward Mono and Inyo Counties for possible new sources of income, and in the spring of 1880 narrow-gauge rails began to be extended southward from Mound House, Nevada, a station on the Virginia and Truckee ten miles east of Carson City. Three years later the line reached the Owens Valley. Although the original intention of the builders had been to connect Carson City with the Colorado River by rail, the widespread economic depression of the 1880s destroyed the market for silver, forcing the railroad to "dead-end" at Keeler (formerly Hawley) on the east shore of Owens Lake.

As with other narrow-gauge lines constructed in the mountain West, the function of the Carson and Colorado was to transport bulky freight over rugged terrain to the main-line railroads, then on to distant markets. To the north of the Owens Valley, its rails were put down across the fragmented topography of the western Great Basin, finally skirting the western flank of the White and Inyo Mountains to tap what remained of the mining business in the region. Since its route bypassed the valley's west-side agricultural communities, each of the valley's permanent settlements had to build a rail depot three to five miles distant on the east side of the Owens River. Laws became the valley's principal station, serving the Bishop–Round Valley region. The Alvord, Citrus, and Mt. Whitney stations served Big Pine, Independence, and Lone Pine, respectively. Ironically, after years of pushing for a southern rail connection – the valley's natural outlet – the first line came in from the north, providing circuitous connections with the San Francisco Bay Area rather than with southern California. Many products of the mines that had formerly been carted to the rail depot serving Los Angeles at Mojave by twenty-mule-team wagons, now began to move north by rail.

But the Owens Valley's distance from the transcontinental Central Pacific Railroad at Reno, and the required transshipment of commodities from narrow- to standard-gauge trains, meant that freight rates were exceedingly high – too high to allow the majority of farm products to be profitably shipped out of the valley. The rail line did broaden the beef market somewhat, making it possible to ship livestock to West Coast markets. But otherwise the Carson and Colorado did little to stimulate the valley's slumping agricultural economy. In fact, in one respect the railroad hurt the valley's farm economy, particularly in the Bishop–Round Valley region. Before it arrived, some farmers could still earn a little income by hauling farm products to the Benton, Bodie, and

Candelaria mining camps to the north, thus enhancing the rather meager cash receipts they received for their produce. The arrival of the Carson and Colorado all but eliminated this supplemental source of revenue.[2]

Finding itself without the anticipated mining activity upon which the Carson and Colorado's livelihood depended, the railroad eventually took steps to encourage immigration to the Owens Valley, particularly to land east of the river. The railroad purchased land near each of its depots in the valley and planted thousands of cuttings of wine, table, and raisin grapes to prove to potential settlers the fertility of the barren, sagebrush-covered east-side lands.[3] Most westward migrants, however, shunned the long and arduous side trip to the Owens Valley. From the transcontinental line at Reno, a prospective settler would have to ride the Virginia and Truckee Railroad to Mound House to connect with the Carson and Colorado, then travel more than 200 miles across sparsely populated desert terrain to reach the northern end of the valley. From the south, the traveler faced a 120-mile trip on a stagecoach that ran semiweekly across the Mojave Desert from the transcontinental rail depot at the town of Mojave to the Carson and Colorado's southern terminus at Keeler. The railroad therefore not only carried little freight but few passengers as well. Ten years after the initiation of service by the Carson and Colorado, the people of the Owens Valley were still pining for a railroad connection with Los Angeles.[4]

The Desert Land Act and the Owens Valley Land Rush

In 1880 most of the public land in the Owens Valley was still open for settlement, for the majority of patented land entries at the time clustered near the numerous streams that flowed off the Sierra (Fig. 7.1). When the *Inyo Independent* reported in 1877 that stream flow off the Sierra Nevada had largely been appropriated, it also noted that "the principal and perfectly reliable and abundant resource for water for all the untaken lands remains untouched. This is Owens river. . . . All the land that Owens river will irrigate is yet open to settlement."[5] Hopeful about pending legislation that would expedite the use of Owens River water, the newspaper added: "Should the bill now pending in Congress to extend the so-called Lassen County Desert Land Law to the benefit of all the States and Territories become a law . . . the inducements will be greatly increased for present residents to undertake a river irrigating enterprise."[6] On March 3, 1877, Congress did in fact extend the provisions of the Lassen County Desert Land Act, though with some modifi-

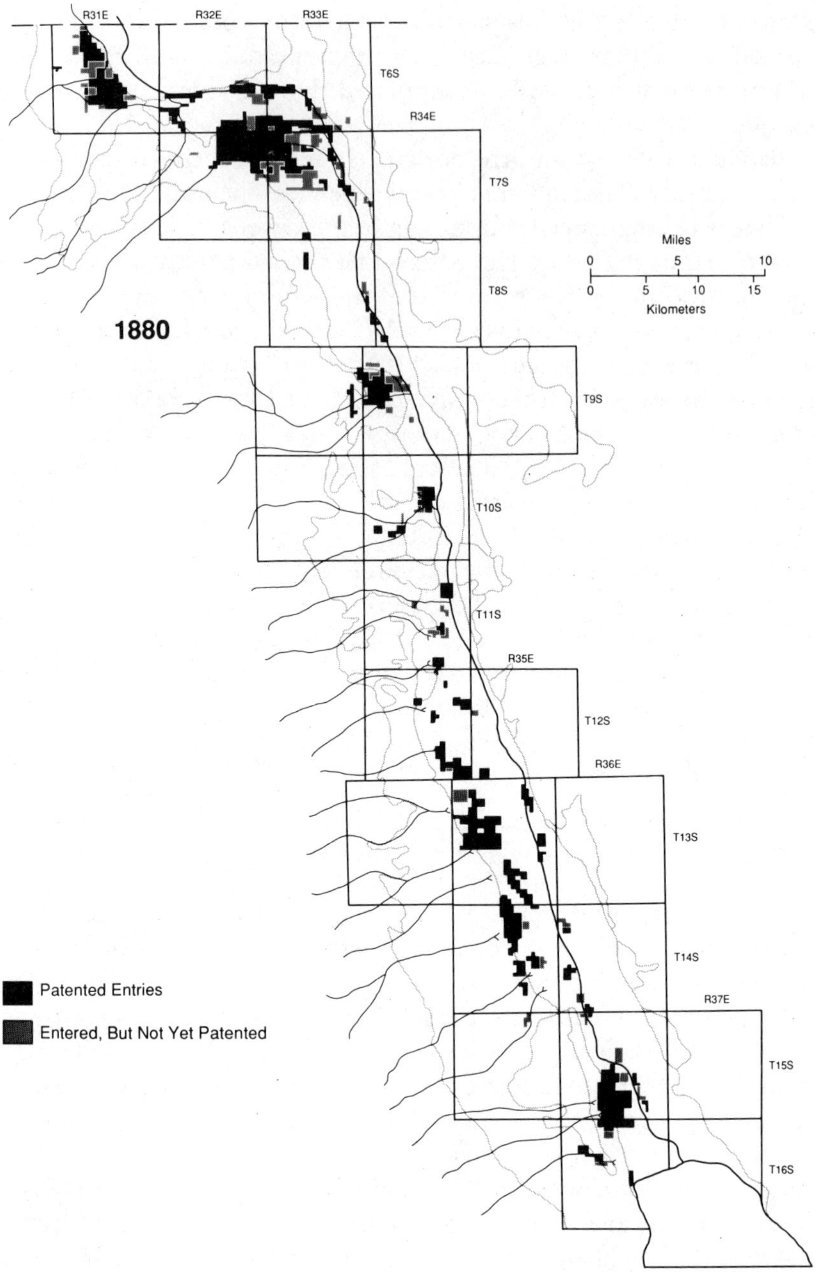

Figure 7.1. Patented entries in the Owens Valley through 1880. (Compiled from Independence District Land Office tract books; reproduced from Sauder, "Patenting An Arid Frontier," 554, by permission of the Association of American Geographers)

cations, to all the Western states and territories except Colorado.[7] In the Owens Valley the *Inyo Independent* proclaimed it "the best land law ever yet enacted," because it would allow potential land entrants to acquire large enough tracts to make the expense of their reclamation worthwhile.[8]

During its first year, twenty-three desert land entries were made in the Owens Valley, though all but two of them were eventually canceled. On April 14 the *Inyo Independent* reported that "John Broder made the first deposit in this Land Office for a section of land under the desert land law, and J. W. McMurry, the second."[9] Both men had put together large cattle herds in the valley, and like other ranchers in the arid West, they initially used the Desert Land Act to control the range.[10] Most of the earliest desert land filings in the Owens Valley were made in its relatively open, poorly watered southern half, where there was a greater reliance on stock raising, and even though proof of reclamation was required by law after three years, many of the entrants held their claims for seven years or more. Cattle interests saw the Desert Land Act as a means of acquiring occupancy or use rights to relatively large blocks of land without the burdens of ownership. Consequently, most early entrants never submitted final proof, because they had made no attempt at reclamation through the development of irrigation. Congress, in formulating a reclamation policy for Western arid lands, had apparently overlooked their potential use for grazing purposes.[11]

Following this initial wave of desert land filings in the Owens Valley, the number of subsequent entries remained relatively insignificant through the early 1880s. Nevertheless, as predicted, passage of the Desert Land Act resulted in the establishment of a number of irrigation canal projects in the valley. Although the Desert Land Act made no specific provision for reclamation except by individual effort, nothing prevented the land entrants from joining together to build canals to irrigate their tracts.[12] As a result, most canals in the Owens Valley were financed on a cooperative basis. Ditch companies were formed and incorporated, and farmers purchased shares of stock in them, with each share carrying the right to use a designated amount of water. The farmers themselves built the canals during the winter using teams of horses and primitive cast-iron scoops called Fresno scrapers. Most of these partnerships built relatively small ditches that irrigated only the lowlands, while the higher alluvial slopes remained unwatered. The McNally Ditch and Fish Slough Canal (two parallel ditches serving the dry land northeast of Bishop), the Bishop Creek Ditch (for tracts between Bishop and the Owens River), and the Big Pine Canal (for the land east of Big Pine) were the first to be projected. "No enterprise yet set foot in this country gives higher promise of profit and lasting good than these irrigating canals," stated the *Inyo Independent*.[13] The McNally Ditch and the Fish Slough Canal, the smallest of the original enterprises, were the first to be completed, and by 1880 the two ditches were "already yielding handsome returns on the investment of labor and money."[14]

By 1885 seven ditches were conducting Owens River water to the arid lands of the valley, and a gradual increase in patented desert land entries – associated with the slow extension of irrigation ditches in the valley's north end – continued into the mid-1880s.[15] Then in the summer of 1886 there began a genuine rush for the Owens Valley's unclaimed public land. By 1890 a large portion of the valley floor was under claim to which title would ultimately be conveyed (Fig. 7.2). Two widely separated irrigation enterprises were largely responsible for this intensified pattern of land alienation. In the north, the Owens River Canal would serve the vacant land west and southwest of Bishop. It was similar to all the others in that its construction involved cooperative effort by local farmers. In the southern townships, Nevada capitalists began construction on the Eastside Canal. It was to water the vacant tracts along the east side of the river from the vicinity of Independence southward.

Since the alignment of both canals crossed large tracts of unclaimed desert land, nonresidents began to speculate in desert land claims based on their expectation of the increased values that would result from them. Even more than the Homestead and Preemption acts, the Desert Land Act provided an opportunity for "dummies" to make land entries and assign them to speculators,[16] and the *Inyo Independent* unintentionally facilitated desert land speculation early in 1886 by publishing the amount of acreage capable of irrigation that remained unclaimed in each township in the valley – a total of nearly 70,000 acres. Figure 5.2 shows the magnitude of this speculative activity when the number of patented and canceled desert land entries made in 1886 and 1887 are compared with those made in previous years.

Initially, the heavy desert land activity was not unwelcome, because local interests naively anticipated that the locators would actively participate in canal construction. The nonresident speculators offered little help with the canal building, however, and their inactivity was quickly denounced by the *Inyo Register*, the new weekly newspaper in Bishop. "Any man who takes up land with no intention of improving it," the editor declared, "is a land-grabber of the meanest sort."[17] Somewhat later the *Independent* also conceded that land specu-

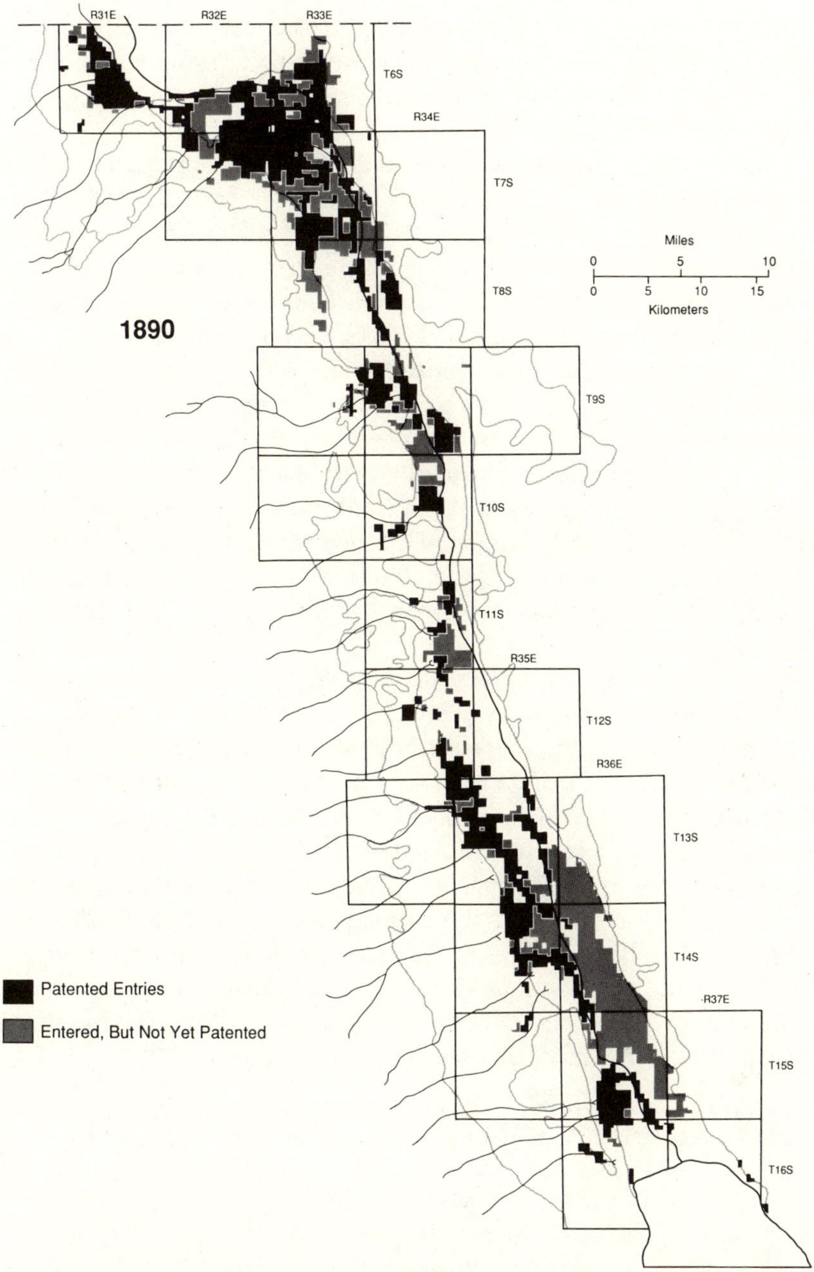

Figure 7.2. Patented entries in the Owens Valley through 1890. (Compiled from Independence District Land Office tract books; reproduced from Sauder, "Patenting An Arid Frontier," 556, by permission of the Association of American Geographers)

lation in the valley was retarding the region's development. "Locating lands and holding them without doing anything to improve them is a hindrance and injury to all settlers in localities where it is practiced," its own editor concluded.[18]

Much of this criticism was directed toward the activity occurring in the vicinity of the Owens River Canal and the Eastside Canal. Two colonization schemes – one fraudulent, the other legitimate – surfaced in the Owens Valley in connection with these two enterprises. The first, calling itself the International Immigrant Union, used dummy land entrants to file on several contiguous tracts west of Bishop along the line of the projected Owens River Canal. A prospective settler would pay the standard $1.25 per acre for a parcel, but the Union assessed an additional fee of twenty-five cents per acre for its aid in placing the settler on the land. Water was to be supplied through the purchase of stock at ten dollars per share in a fallacious enterprise known as the Owens Valley Canal and Water Company.[19] Since no effort was made to get water to the lands located by the Union, all of its entries were subsequently canceled, leaving those who had purchased parcels from the firm landless.[20]

A more sincere but equally unsuccessful operation involved the Eastside Canal. More than twenty thousand acres were entered under the Desert Land Act in the southern half of the valley by a Nevada syndicate known as the Inyo Land Company. Land agents were hired to file contiguous desert land entries east of the Owens River from townships 13 through 15 south, and a sophisticated ditching machine that had been brought from Chicago and that employed three men and sixteen horses was used to dig the Eastside (Inyo) Canal to irrigate the lands.[21] By October 1888 water was being diverted into the first ten miles of canal. The *Inyo Independent,* still optimistic about the impact of the Desert Land Act on the valley's future, lavished the following praise on the activities of the Inyo Land Company:

> In marked contrast to others who located lands in Owens Valley about the same time is the action of parties who located land on the east side of the Owens River from Independence southward. These men went to work in good faith and are building one of the finest irrigating canals in the State. This enterprise has already made a large amount of taxable property; in addition to many thousands of dollars of hard coin expended amongst us for labor and supplies of all kinds. These are the speculators who bring prosperity to a country; they develop its resources and encourage other enterprises. A very few years more, will see twenty thousand acres or more of vineyards, orchards and green

> fields, and hundreds of comfortable homes, along the line of the Eastside Canal and this grand result will be due to the men of Virginia, Carson and Reno. But for these men that great tract would have lain for untold years a sagebrush desert.[22]

Two years later, however, the newspaper's optimism had turned to disappointment: "Several thousand acres have already been proved up on, but so far, no land under the Eastside Canal has been put under cultivation. . . . Last summer the locator quit further efforts to do anything with the place, and the drifting sands have since piled up on it."[23] The editor also noted that "from Independence to Owens Lake, a distance of twenty-one miles, and including the best fruit land in the valley, during the past six years but four families have located and built houses, and of these four only one attempts to make a living from his farm."[24] Hence, even when water was brought to the land in good faith by desert land locators, there was no guarantee of success. The Owens Valley was still suffering from a severe economic slump, and settlement would progress only slowly.

The Timber Culture Act likewise facilitated land speculation in the Owens Valley in response to the extension of irrigation canals. Although far fewer timber culture entries were filed than desert land claims, a much larger proportion of tree claims was ultimately canceled (Fig. 5.2). As with desert land filings, the majority of the timber claims were made by nonresidents who expected to hold their claims until water in one of the projected canals would boost the value of their land. Also, homesteading (and to a lesser extent preemption until its repeal in 1891) was rejuvenated in the Owens Valley as a result of the new irrigation projects (Fig. 5.2). Undoubtedly, some of these homestead locations were made for speculative purposes, for the incidence of homestead cancellations was somewhat higher than in previous years. But the higher cancellation rate may also have been due both to the relatively slow pace at which the irrigation canals were put through and to the depressed economic conditions of the time. Homesteading and preemption nonetheless again emerged as viable methods of land acquisition once water from the irrigation ditches became available for crops.

Since nonresident locators of desert lands provided little assistance in putting through the ditches, the extension of the majority of canals in the Owens Valley continued to progress slowly. By the early 1890s only fifty-seven miles of canals were in operation, and although twenty-one ditch enterprises totaling 250 miles in length had been projected, very few canals had been diverting Owens River water long enough to give returns commensurate with the amount of labor and money they had cost.[25]

Nevertheless, by 1901 water from the Owens River was being diverted through eighteen main ditches and canals and some smaller ones to create an irrigation system nearly 200 miles long.[26] Most were located in the northern end of the valley because the larger number of farmers and the more compact form of settlement there made construction on a cooperative basis more feasible compared with the widely dispersed pattern of settlement of the southern region. As a result, 63 percent of all desert land entries made in the valley's northern townships were carried to patent.[27] In contrast, south of Poverty Hills only 34 percent of the desert land entries reached the final patent stage, a figure equal to the average for the entire West.[28] The expanding irrigation works in the Bishop region gave desert land locators in that section a significant advantage over those who had located land in the southern townships.

Perhaps the primary reason that the work of canal construction continued despite the prevailing economic recession was the belief among Owens Valley residents that if a southern extension of the railroad were ever to become a reality, they needed to enhance their agricultural status. The *Register* encouraged work on the canals because, its editor believed, "without additional irrigating canals, there is room here for but a few new comers."[29] More settlers would hasten the process of canal construction, and this in turn would eventually result in farm production in excess of home demand. The subsequent abundance of freight presumably would lead to the construction of a railroad extension to Mojave, thereby allowing the surplus produce to flow unimpeded into southern California. Unfortunately, the slow progress on the canals attracted few new migrants to the Owens Valley, nor did the canals result in a dramatic increase in the valley's agricultural output. The empty lands between Owens Lake and Mojave remained unspanned by a railroad.

Farm Enlargement

Finding themselves seemingly forever saddled with excessive freight rates to West Coast markets, Owens Valley farmers began to distance themselves even further from the mixed farming system that they had originally established in the valley. Instead, they looked to alternate means of producing something that could profitably make the long journey out of the valley. Stock raising had always been a major

Table 7.1. Number of Farms, Farm Acreage, and Improved Acreage in Inyo County, 1880–1900[a]

	1880	1890	1900
Number of farms	242	225	424
Average farm size (acres)	209	303	333
Acres in farms	50,487	68,256	141,059
Improved (irrigated) acres	28,323	46,242	41,026
Improved acres as proportion of total farm acres	56%	68%	29%

SOURCES: U.S., *Tenth Census, 1880, Report Upon the Statistics of Agriculture*; U.S., *Eleventh Census, 1890, Report of the Statistics of Agriculture*; U.S., *Twelfth Census, 1901*, Agriculture, Part I.

[a]Since detailed federal manuscript census schedules are not available for Inyo County after 1880, the published figures represented here encompass all of Inyo County rather than only the Owens Valley. Farm statistics for the Owens Valley, however, are essentially those for the county, because nearly all the farming in the county was done in the valley.

activity, and with consumer demand for beef expanding, it took on increased importance, particularly since livestock was about the only Owens Valley commodity that could justify the cost of rail shipment to West Coast markets. Within a year after the Carson and Colorado entered the valley, one contemporary observer of farm conditions in the Owens Valley noted that "nothing the farmer produces is as easily converted into money as beef."[30] The extensive range peripheral to the valley, combined with the ease and cheapness with which superior crops of alfalfa could be grown within the valley, facilitated the expansion of the cattle industry. With livestock raising destined to become the leading economic activity, the characteristic 160-acre farm unit began to be threatened. The small farmers' inability to market grain and other products of general farming, which caused an outmigration of many of the valley's pioneer farmers, and the tendency of stock raising to concentrate in the hands of a few individuals, together facilitated the absorption of quarter-section parcels into large cattle ranches. Even though acreage in farms increased significantly during the 1880s, the federal census of 1890 recorded a decrease in the number of farms in Inyo County, while the average farm size increased by nearly one hundred acres (Table 7.1). Now, 12 percent of the valley's farms were 500

Table 7.2. Number of Farms by Farm Size in Inyo County, 1880–1900

	1880		1890		1900	
	No.	%	No.	%	No.	%
Less than 50 acres	18	7.4	11	4.9	90	21.2
50 to 99 acres	22	9.1	20	8.9	40	9.4
100 to 499 acres	192	79.3	167	74.2	248	58.5
500 to 999 acres	5	2.1	15	6.7	28	6.6
1,000 acres or more	5	2.1	12	5.3	18	4.2

SOURCES: U.S., *Tenth Census, 1880, Report Upon the Statistics of Agriculture*; U.S., *Eleventh Census, 1890, Report of the Statistics of Agriculture*; U.S., *Twelfth Census, 1900, Agriculture*, pt. 1.

acres or more in size, compared with only 4 percent ten years earlier (Table 7.2).

The trend toward bigger farms was not enthusiastically endorsed by the valley's newspapers, particularly in localities south of Big Pine, where the process was most pronounced. The southern region was better suited to stock raising because of its less abundant natural water supply, smaller number of irrigation canals compared to the north, and much smaller home market for local produce. The *Inyo Independent* maintained in 1887 that "while it is very desirable to have capital invested in the county, yet if that capital is to be used in buying out small farmers in order to make big stock ranches, we would be better without it than with it."[31] Later, in reflecting on the high freight rates for shipping farm products out of the valley, the newspaper recommended growing fruit as an alternative to stock raising.[32] Although fruit production, particularly of apples and peaches, was of growing importance, it had not yet expanded beyond the local demand.

Specialization in some profitable activity to forestall the expansion of stock raising also concerned northern valley residents. The *Inyo Register*, however, expressed reservations about relinquishing the diversity of production that had characterized the region in the past:

> Exclusiveness to one line of crop is not to be recommended where diversity is possible. . . . We doubt if there is any section of country from the Columbia to the Colorado that will successfully produce as many

> different kinds of crops as does this Owens River Valley. This is very fortunate for our local needs, which has been about the only market open to us so far. But the time is at hand when we will have to look elsewhere for a market, not, however, for our products in general – we must look to the production of some leading article that will bear transportation, and there are but very few indeed that will do that unless offered in quite large quantities, and some not even then under existing conditions – grain, for instance. As yet this valley . . . could not safely undertake to regularly supply any outside market of consequence with any shippable product, beef probably excepted. As a feasible specialty the creamery proposition promises to work out more important commercial results of a fixed and valuable character for the least outlay than any other one industry as yet within our means.[33]

The systematic development of the Owens Valley's dairy industry began in the northern end of the valley in 1892 when the Inyo Creamery was incorporated and a plant was constructed at Bishop.[34] Yet, although both fruit growing and dairying would eventually gain in importance, neither would be instrumental in slowing the expansion of stock raising.

According to the 1890 census, the number of acres devoted to cereal production in the Owens Valley had declined since the previous census, and the output of grain had likewise decreased by 24,000 bushels, largely as a result of the steep decline in the production of barley (Table 7.3). With most of the mining centers of the region closed, the primary market for barley had dissolved. Only wheat registered a significant increase in both acreage and output. It became the valley's principal

Table 7.3. Grain Production in Inyo County, 1880–1900

	1880		1890		1900	
	(Bu.)	(Acres)	(Bu.)	(Acres)	(Bu.)	(Acres)
Corn	33,213	1,682	26,759	1,093	53,076	2,106
Wheat	30,004	1,525	36,548	1,963	63,293	2,249
Oats	22,538	791	24,716	794	51,900	1,154
Barley	35,845	1,686	9,530	361	7,630	195
Total	121,600	5,684	97,553	4,211	175,899	5,704

SOURCES: U.S., *Tenth Census, 1880, Report Upon the Statistics of Agriculture*; U.S., *Eleventh Census, 1890, Report of the Statistics of Agriculture*; U.S., *Twelfth Census, 1900*, Agriculture, pt. 2.

Table 7.4. Hay and Livestock Production in Inyo County, 1880–1900

	1880	1890	1990
Hay (tons)	8,417	24,574	37,247
(acres)	5,346	9,550	14,723
Horses (no.)	3,287	4,033	4,481
Milk Cows (no.)	1,273	1,075	1,575
Range Cattle (no.)	5,997	12,447	11,119
Sheep (no.)	9,574	12,682	62,036
Wool (lbs.)	35,382	139,252	264,290
Swine (no.)	2,672	1,758	2,588

SOURCES: U.S., *Tenth Census, 1880, Report Upon the Statistics of Agriculture*; U.S., *Eleventh Census, 1890, Report of the Statistics of Agriculture*; U.S., *Twelfth Census, 1900, Agriculture,* pts. 1 and 2.

cash crop and was grown in rotation with corn. Oats, another important product of the valley, was also rotated with corn. The reduction in the valley's grain harvest was offset by a threefold increase in hay production and a near doubling of the acreage devoted to hay, reflecting the expansion of the valley's livestock industry (Table 7.4). The increased number of range cattle and sheep in the county, along with increased wool production, also confirm the heightened importance of stock raising. Milk, eggs, butter, and honey production, most of which served the home market, all registered gains, but the value of orchard and market garden production had declined markedly (Table 7.5).

Economic Stagnation

The decline in the valley's grain harvest reflected the difficulties associated with general farming in the Owens Valley. One farmer, reportedly among the best in the entire valley, averaged an annual income of only a hundred dollars from the products raised on his 240-acre farm during most of the 1880s.[35] Another valley resident, a pioneer who had been farming in the region for twenty-five years, expressed his distress over the stagnating economy:

> There is no money in farming. Recently I took a load of various products, flour, hams, bacon, and other articles, hauled them forty miles and offered the goods to merchants. The load should have realized several hundred dollars, but all the coin I got was $27. The balance I took in exchange of goods of various kinds, most of which I could have done without.[36]

A third farmer, who owned four hundred acres of choice land in the valley, summarized the forces operating against the general farmer in the Owens Valley:

> It is almost impossible to export farm produce from the valley. Freight cost is a complete bar to that. The home market is so limited that but a small amount of produce can be consumed. What I expect for the future is, that as stockmen have lately been increasing their interests in the valley, this will grow till stock raising will be the only interest left of any importance, and the land will be owned mostly by stockmen.[37]

Table 7.5. Miscellaneous Farm Products in Inyo County, 1880–1900

	1880		1890		1900	
	(Trees)	(Bu.)	(Trees)	(Bu.)	(Trees)	(Bu.)
Apples	n.a.		5,393	3,438	28,571	6,413
Peaches	n.a.		4,823	3,306	9,041	1,254
Pears	n.a.		313	325	3,313	893

	1880	1890	1900
Milk (gal.)	510	223,305	464,633
Butter (lbs.)	44,395	55,316	47,813
Eggs (dozen)	22,660	47,813	122,730
Honey (lbs.)	10,630	109,665	146,320
Market Garden/ Orchard Prods. ($)	10,403	1,992	27,232

SOURCES: U.S., *Tenth Census, 1880, Report Upon the Statistics of Agriculture*; U.S., *Eleventh Census, 1890, Report of the Statistics of Agriculture*; U.S., *Twelfth Census, 1900, Agriculture*, pts. 1 and 2.

The latter observation would ultimately prove true, at least for the central and southern portions of the valley. In 1896 Thomas Rickey, a Nevada cattle baron, began to expand his cattle domain in the area south of Poverty Hills by requesting state lieu land selections. As mentioned earlier, when a prospective purchaser such as Rickey made selections of unpatented lands at the state land office, the state would acquire federal patents to these lands as lieu selections. Although parcel size was limited by law to 640 acres per locator, this proviso was easily sidestepped by employing agents to make the desired land entries. In this manner, Rickey entered extensive state lieu land acreages in two townships, claiming much of the remaining unappropriated lands in the valley. Consequently, by the turn of the century the process of land alienation on the valley floor was nearly complete (Fig. 7.3). Most of Rickey's original selections consisted of a marshy salt grass meadow bordering the Owens River that was of little use to farmers. But his lieu land filings were the catalyst for the development of a cattle domain in the Owens Valley that would eventually encompass more than 20,000 acres. Because of the gloomy economic prospects, many of the region's general farmers would eventually sell their holdings to Rickey.

Toward the end of the 1890s, little had happened in the Owens Valley to counter the economic trends initiated two decades earlier. In 1897, in its annual survey of California counties, the *San Francisco Chronicle* confirmed the valley's ongoing difficulties: "Owing to its isolated location Inyo County has fallen far behind her sister counties in the marvelous development that has characterized California in the past ten years. [Because of] limited facilities for transportation, little grain is raised beyond that needed for home consumption, and that, indeed, is true of all agricultural products."[38] The *Chronicle,* however, did underscore the extensive valley acreage devoted to alfalfa, "which yields immense crops."

Landscape Improvements

Because of the Owens Valley's economic difficulties, Inyo County's population grew slowly during the last two decades of the nineteenth century, increasing by only 616 persons in the 1880s and another 833 in the 1890s.[39] Yet in spite of slumping economic conditions, landscape alterations in the valley continued at a steady pace. By the mid-1880s there was a noticeable increase in "trees, crops, and verdure" as a result of the expanded irrigation network.[40] Bishop had sustained its dominance as the leading agricultural community in the valley, and its more

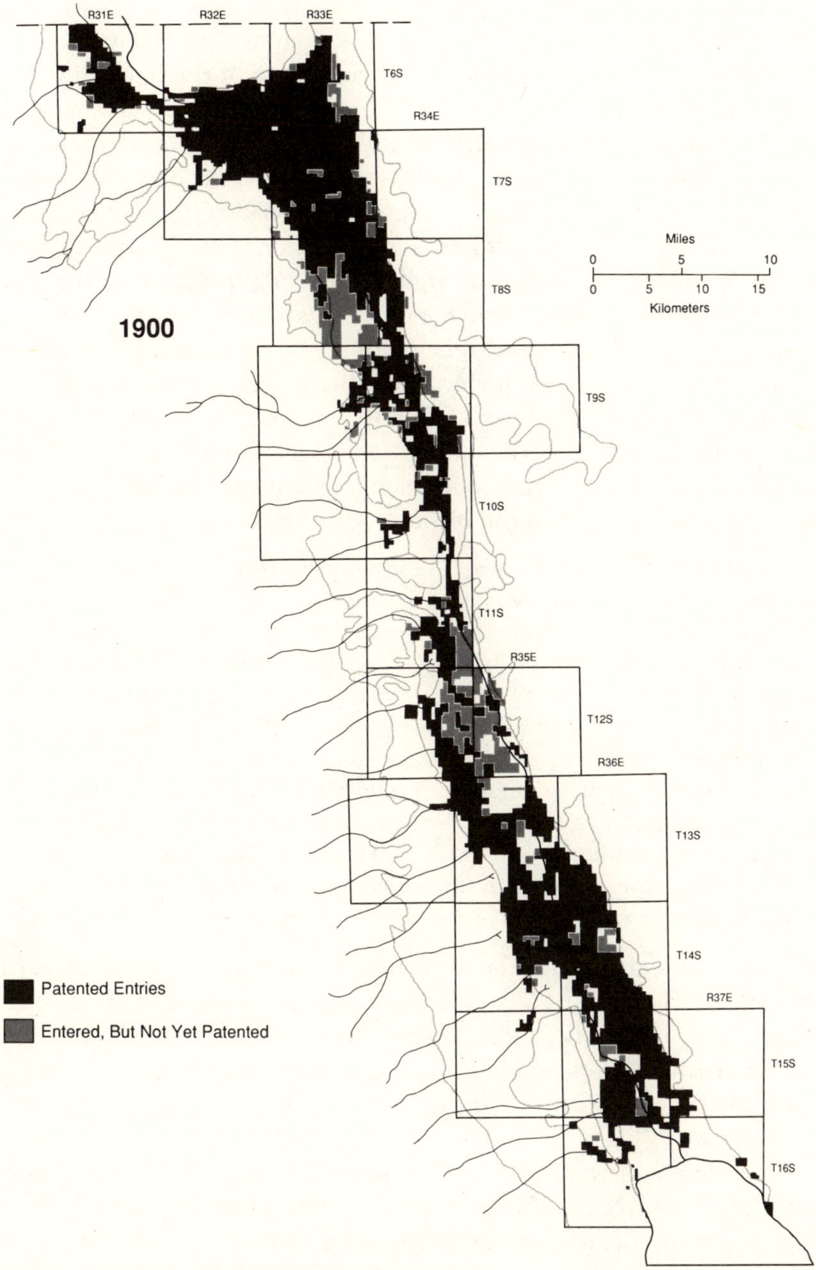

Figure 7.3. Patented entries in the Owens Valley through 1900. (Compiled from Independence District Land Office tract books; reproduced from Sauder, "Patenting An Arid Frontier," 560, by permission of the Association of American Geographers)

settled appearance, "where long rows of . . . trees now form excellent shade . . . and add much to the beauty of the landscape," as well as the ongoing work of canal extension, attracted most of the new migrants who came into the valley.[41] Still, in the mid-1880s thousands of acres of fertile land were open for settlement in the valley's northern end east of the Owens River.[42] Sagebrush-covered land could be cleared relatively easily with a team of horses and a homemade grubbing machine at an average cost of 50 cents an acre,[43] and improved land could be purchased at prices ranging from eight to thirty dollars per acre, depending on the type of improvements and the location of the parcel.[44]

As the progress of canal extension in the northern end of the valley brought increased acreage under the plow, it was suggested that "no more pleasant view of a farming country can be found" than around Bishop when in late May the plains were covered with fields of alfalfa and grain.[45] In 1893 the *Inyo Independent* reported on the improvements then being made in the Bishop region:

> On all sides land is being cleared of sagebrush, enclosed, sown with alfalfa, wheat, oats, barley or corn or seeded with other crops. Shade trees are planted there more than any place else in the county, and already long stretches of well shaded roads make lovely drives. Only a few years ago that great plain was a brown desert, dreary and barren. Now the eye ranges over a wide expanse of alfalfa fields, the beautiful bright green being a delight to look upon in contrast with the land still unenclosed and in its original desert state. Wisely did the settlers plant trees extensively; and now . . . the whole country for several miles west, north and south of Bishop is richly wooded. Without those groves of trees, no matter how highly cultivated the land might be, the country would still retain much of the original desert aspect.[46]

At about the same time, the *Inyo Register* noted: "In spite of hard times, the area of fenced and cultivated land is continually increasing. . . . It inspires new hope and confidence to get out occasionally and see new land being cleared and land that was sagebrush-covered but a year or two ago now showing alfalfa or other crops enclosed by long lines of young trees."[47]

Improvements were also underway in Round Valley, where about twenty families lived in the early 1890s.[48] With plentiful water derived from mountain streams, no extensive irrigation canals were required for crop production. Some pioneer farmers had by this time encircled their fields with substantial stone walls, and in the northwest corner of the valley sagebrush reportedly "was flying furiously in the air."[49] In

1898 the Owens Valley's first silo was constructed in Round Valley, the first experiment with ensilage in the entire region and one about which dairymen in the northern section of the valley were anxious to learn results.[50] Between Round Valley and Bishop the Hillside Water Company, having acquired the former International Immigrant Union lands, had twenty-five to thirty thousand rooted vines growing on its ranch by 1890.[51]

Southeast of Bishop the settlement of Warm Springs was taking shape. It was watered by Bishop Creek Ditch, where, by the mid-1890s, more improvements were ostensibly taking place than in any other section of the valley.[52] Farther south along the east side of the river, Gustav Sanger (the largest individual landowner in the valley, with a tract comprising about four thousand acres) was growing alfalfa, wheat, and barley on newly cleared land.[53] Toward the end of the 1890s the number of improved ranches and acres of settled territory on the east side of the river were said to be constantly growing.[54] At the same time, the area between Fish Springs and Big Pine had largely been fenced and was rapidly being brought under cultivation: "Forest trees and orchards have been planted. . . . Large fields of alfalfa stretch away."[55] Several farmhouses now dotted a landscape that a few years earlier had been covered with sagebrush. Between Big Pine and Bishop, sixty square miles of land were now enclosed, and much of it was under cultivation.[56]

Nothing comparable to these developments was occurring south of Poverty Hills. "Improvements at any part of Owens Valley," the *Inyo Independent* somewhat defensively remarked, "benefits the whole; and the people in the south end of the Valley . . . [support] their steady advance."[57] After completing its end-of-the-decade survey of the valley's developments, the *Independent* related that "the great object lesson [of the improvements in the north] is irrigation. . . . [U]ntil water is obtained we cannot look for much improvement."[58] The many years of patient labor to construct the canals in the north were now paying off, and the *Independent* advised that "the same intelligent and untiring effort will make just as fruitful a country of the region south of there."[59] With few irrigation canals, little in the way of significant change could be reported for the southern half of the valley. On the vast open plain that later became the cattle domain of Thomas Rickey, for example, there was not a settler to be found in the early 1890s.[60] Beyond this tract and east of the river, thousands of acres owned by the Inyo Land Company were open for settlement, yet only one person was attempting cultivation.[61]

The Evils of Irrigation

Ironically, despite the continued expansion of irrigation works and the associated improvements in the Owens Valley, the amount of irrigated farmland in Inyo County actually declined by 11 percent, or approximately five thousand acres, during the last decade of the nineteenth century (Table 7.1). By 1904 the irrigated acreage in the valley had dropped even further, to an estimated 35,000 acres.[62] Presumably, as new lands were being brought into cultivation, older lands that had been continuously cultivated for many years were simultaneously being taken out of production. For example, in 1898 it was reported that because Round Valley's "hitherto fertile soil has not, of late years, responded so generously as years ago, our salvation and future prosperity lies in the fact of better farming."[63] In light of the rather provincial methods employed by Owens Valley farmers, the *Inyo Independent* commented that a further advantage of a rail link to the south, besides the ability to market produce in southern California, would be to "bring people here that in many ways would make the present residents wake-up to more modern ways of ranching."[64]

The principal problems associated with the farming methods employed in the Owens Valley involved irrigation and drainage. Irrigation simply involved flowing water downslope through a random network of field ditches without control checks or levees, not unlike the methods that had been used by the Owens Valley Paiute (Fig. 7.4). The nature of the problem was summarized in a 1905 survey of agricultural conditions in Inyo County:

> The great majority of farmers in the valley are very lax in their methods. They have taken no special pains to level their land, but irrigate by a system of flooding which over-irrigates in some spots while other spots in the same field may not be touched. This system of flooding has resulted in many of the lower lands becoming too strongly alkaline for any plants other than salt grass, or if wet enough wire grass. In many instances whole fields had been turned into quagmires, and are unfit for anything but grazing purposes.[65]

Not only was too much water voluntarily applied to the fields, but the method of distribution was also wasteful, as seepage from unlined irrigation ditches was substantial. Under such conditions, farmers in the valley found it cheaper and easier to pasture livestock on waterlogged old land and take up new land rather than to improve the system of irrigation.

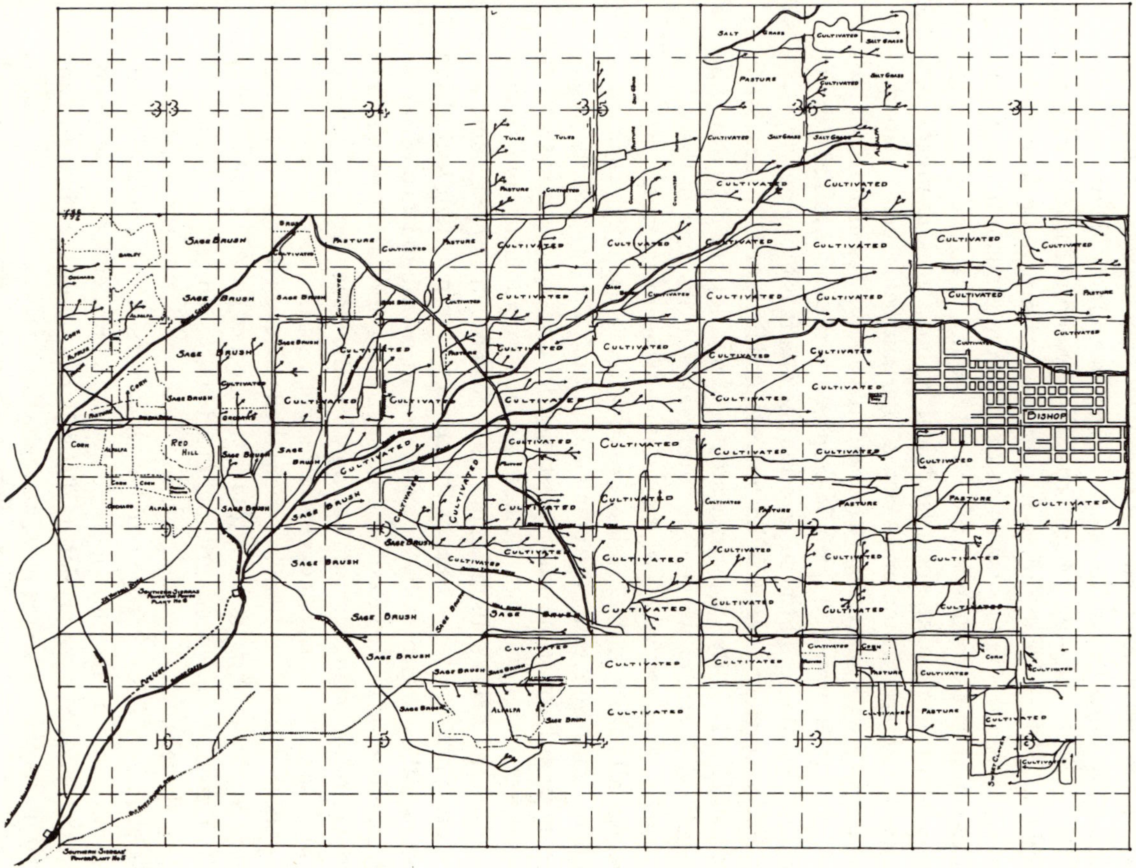

Figure 7.4. The network of irrigation canals and ditches in the West Bishop district represents the haphazard development of irrigation in the Owens Valley. (Courtesy of the Los Angeles Department of Water and Power)

After a quarter century of farming in the Owens Valley, the *Inyo Independent* ran a story on the evils of irrigation when practiced without proper drainage.[66] The newspaper cited the findings of Professor Eugene Hilgard, a noted University of California soil scientist, who found that if irrigation continued from year to year without drainage in areas containing even small amounts of alkali, evaporation would bring the salts to the surface, forming a hard crust that would eventually ruin the land. On level land, continuous irrigation over time would accumulate and saturate the soil to such an extent that crops could not be raised. Focusing on the problem in the Owens Valley, the *Independent* disclosed that "some of the best land near Bishop Creek is already very seriously damaged by too much water. This land before being cultivated was covered with sagebrush; when cleared alfalfa was planted, and for several years it produced splendid crops. Now the soil is full of water to within a few inches of the surface and the alfalfa is dying out."[67]

Unfortunately, Owens Valley farmers paid little heed to these warnings, and by the early 1900s large areas of the valley had been ruined by the excessive use of water. One seventeen-year resident of the valley claimed, "Many ranches which produced the finest crops are now swamp, all owing to the careless use of water."[68] Two periods of irrigation a month were generally employed in the valley, and by the end of the six-month irrigation season, lasting from mid-April to mid-October, an average of 7.5 acre-feet had been applied to cropland.[69] On sandier, more porous valley soils, an application of as much as 15 acre-feet was not uncommon. By the turn of the century, the northern half of the Owens Valley was literally drowning in an overabundance of water: thousands of acres once covered by sagebrush and later brought into production through irrigation had now become either too wet for use or worthless due to surface concentrations of alkali.[70]

Another important environmental effect resulted from this profligate use of water. In the early 1890s the level of Owens Lake began to fall. The *Inyo Independent* announced in November 1894 that the lake's shoreline had receded two hundred feet during the season.[71] In 1898 the *Inyo Register* reported that "hundreds of yards of dry land are now seen around the edge of Owens Lake. . . . Keeler used to be almost at the edge of the water, but its distance from the shore is increasing very noticeably. It seems to be an undisputed fact . . . that [the lake] is diminishing in size."[72] The following year the situation worsened: "The rapid recession of Owens Lake . . . is most observable on the east side. The old Cartago wharf, which in the early '70s stood in eight to ten feet of water, is now about two miles from the present shore."[73]

According to longtime valley residents, the lake level had fluctuated little until 1894. By that time, however, many of the irrigation canals in the northern end of the valley had been put into service, and during the next ten years the shoreline of Owens Lake receded five thousand feet while its level fell fifteen feet, reducing its surface area from 110 square miles to 75.[74] "Buildings which a few years ago stood at the water's edge are now far from it," reported the *Register* in the summer of 1904, "and the old steamer landings are inland, high and dry."[75] The lake level was now dropping at an average rate of 2.5 feet per year.[76] The extravagant irrigation practices of Owens Valley farmers were not only ruining the land, they were also amplifying evaporation and thus reducing stream flow into Owens Lake. The *Register*, in an article entitled "Is Owens Lake Near Its End?" speculated that if the lake level continued to diminish at the present rate, it would be "but 15 years until Owens Lake will have become but a memory."[77] Although the diversion of the surplus flow of the Owens River into the Los Angeles Aqueduct in 1913 hastened the process of reducing Owens Lake to dry alkali flats, the process was already well underway, initiated by Owens Valley farmers themselves.

The Owens Valley at the Turn of the Century

In its annual survey of developments in California counties, the *San Francisco Chronicle* at the turn of the century summarized Inyo County's progress:

> There has been no special movement in Inyo County during the past year to call it directly to the attention of the outside world. The progress has been of the steady, quiet kind, which improves a county without comment. There has been considerable immigration to the Bishop region, that being the best watered section in the county. Several thousand acres of new land have been brought under cultivation, and a number of old farms have passed into the hands of people who will improve them.[78]

The slow but steady progress in the area surrounding Bishop seemed to indicate that people believed that sooner or later the Owens Valley would snap out of its economic doldrums. In 1900 the Southern Pacific purchased the Carson and Colorado Railroad, and this inspired new hope for the long-awaited southern extension to Mojave. But rail service leading south did not come until a few years later, and then it was a by-product of the construction of the controversial Los Angeles Aqueduct.

Instead, in the summer of 1900 a Nevada mining boom at Tonopah and Goldfield set the stage for a period of renewed economic development in the valley.

The discoveries at Tonopah and Goldfield, both of which were little more than a hundred miles from Bishop, opened a new and lucrative market for farm products from the Owens Valley. By early 1903 the economic recovery that was so long in coming to the valley was well underway: "The year which has just joined the receding procession has been one of greater prosperity for this county than most of its predecessors. Ready market has been at hand for everything that could be sold," reported the *Register.*[79] By 1904 Tonopah had an estimated population of eight thousand (plus four thousand horses), and nearly all of the mining camp's supplies came from the Owens Valley.[80]

Agricultural census figures for Inyo County for the federal census of 1900 provide some indication of conditions prevailing in the valley on the eve of the Nevada mining boom. One striking feature is the almost doubling of the number of farms since the previous decennial census, accompanied by more than a doubling in farm acreage (Table 7.1). Concurrently, however, the proportion of improved acreage dropped precipitously. This largely reflects the patenting of most of the dry lands on the valley floor in the 1890s by desert land entrants who participated in the Owens Valley's land rush but who had done little to reclaim and improve their landholdings.

Overall, the tendency toward larger farms continued but at a slower pace than in previous decades, with the average farm size increasing from 303 acres in 1890 to 333 in 1900. Hidden in the latter figure, however, were two divergent trends now underway in the valley. Because of the continuing work on the irrigation canals, the majority of acreage devoted to irrigated farming in the valley was concentrated in the northern end (Fig. 7.5). There, the improved irrigation facilities in the north were used to expand alfalfa production for the valley's dairy and livestock interests. Both hay output and the number of acres devoted to hay production increased significantly during the 1890s, as did wool production and the number of milk cows and sheep (Table 7.4). But the steady advance of irrigation in the north also resulted in the subdivision of many farms into smaller units (Fig. 7.5). By 1900, 130 of Inyo County's 424 farms, or 30 percent, were less than 100 acres in size, and the majority of these farms encompassed fewer than 50 acres (Table 7.2).

The expansion of irrigation works and the move toward smaller farms in the northern end tempted some farmers and ranchers to exper-

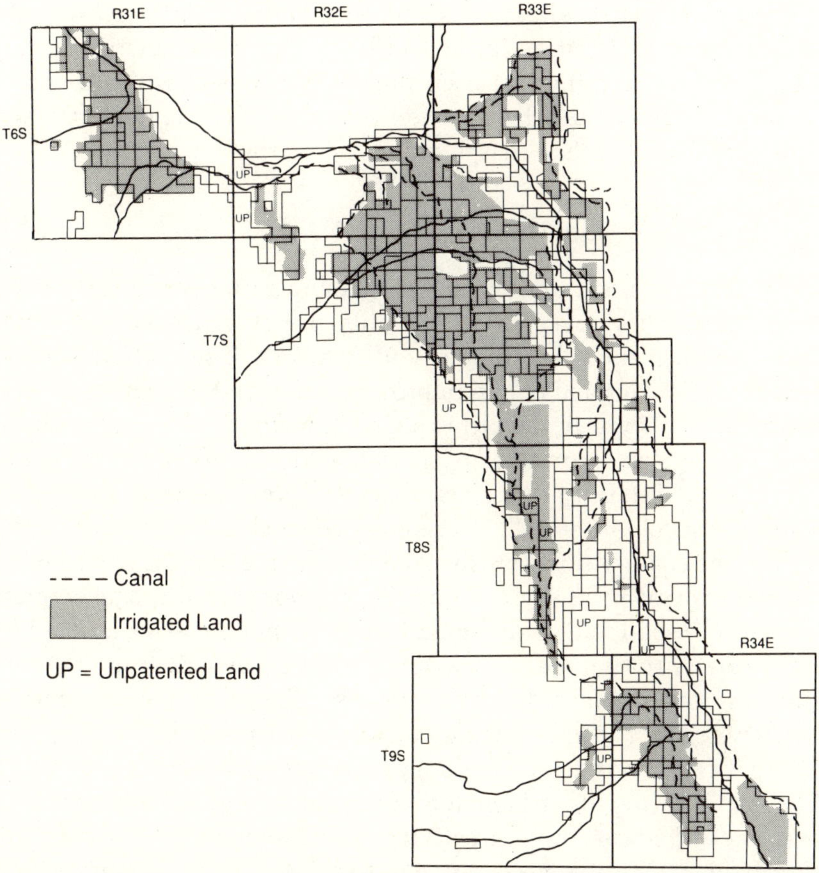

Figure 7.5. Irrigated land and land ownership boundaries in the Owens Valley's northern townships, 1905. (From property assessment maps, Inyo County Assessor; reproduced from Sauder, "Powell's Vision of Arid Land Settlement Reexamined," 85, by permission of the Association of Pacific Coast Geographers)

iment with more varied activities, such as fruit production, dairying, and poultry raising.[81] Perhaps the most striking change at the turn of the century is reflected in the number of fruit trees planted during the previous decade, along with the expansion of milk, egg, and honey production (Table 7.5). The value of market garden and orchard products, rising from approximately $2,000 to over $27,000, also reflects the valley's new economic fortune. The newly irrigated areas and the expanded demand by the mines also helped restore the importance of the traditional staples, particularly wheat, oats, and corn (Table 7.3). In contrast, the development of irrigation in the south was much less impressive (Fig. 7.6). Intensive farming had been unable to compete with expanding livestock interests, and the average size of individual landholdings was considerably larger than that found in the north.

North Versus South

A closer examination of land ownership patterns and early twentieth century improvements highlights the evolving regional differences between the northern and southern ends of the valley. The pattern of settlement and public improvements found in a representative township (T7S R33E) in the early 1900s attests to the cooperative efforts of the valley's northern settlers, as well as the ability of the prevailing public land system to foster permanent improvements in this better-watered portion of the valley. The grass-covered fine sandy loams adjacent to the copious flow of Bishop Creek were the first lands to be taken up in the township (Fig. 7.7A). Initially, the homestead and preemption laws were used to alienate public land here, but a few desert land entries were also patented. By 1900 most of the valley floor in the township was under patented entry (Fig. 7.7B). The primary reason for the land entry activity of the late nineteenth century was the ongoing work of canal construction, and several canals now traversed the township. The initial cluster of settlement that resulted spontaneously from early homestead and preemption activity near Bishop Creek provided the threshold population necessary for the cooperative construction of canals to carry water to heretofore dry tracts of land, and the cooperative irrigation canals in turn had a significant influence on the subsequent progress of settlement and improvement in this township (Fig. 7.7C).

The situation was different, however, south of Poverty Hills. The region near Lone Pine, as exemplified in another township (T15S R36E), was significantly drier than the northern end of the valley. The majority of this township's earliest inhabitants, filing under the Preemption Act,

Figure 7.6. Irrigated land and land ownership boundaries in the Owens Valley's southern townships, 1905. (From property assessment maps, 1905, Inyo County Assessor; reproduced from Sauder, "Powell's Vision of Arid Land Settlement Reexamined," 86, by permission of the Association of Pacific Coast Geographers)

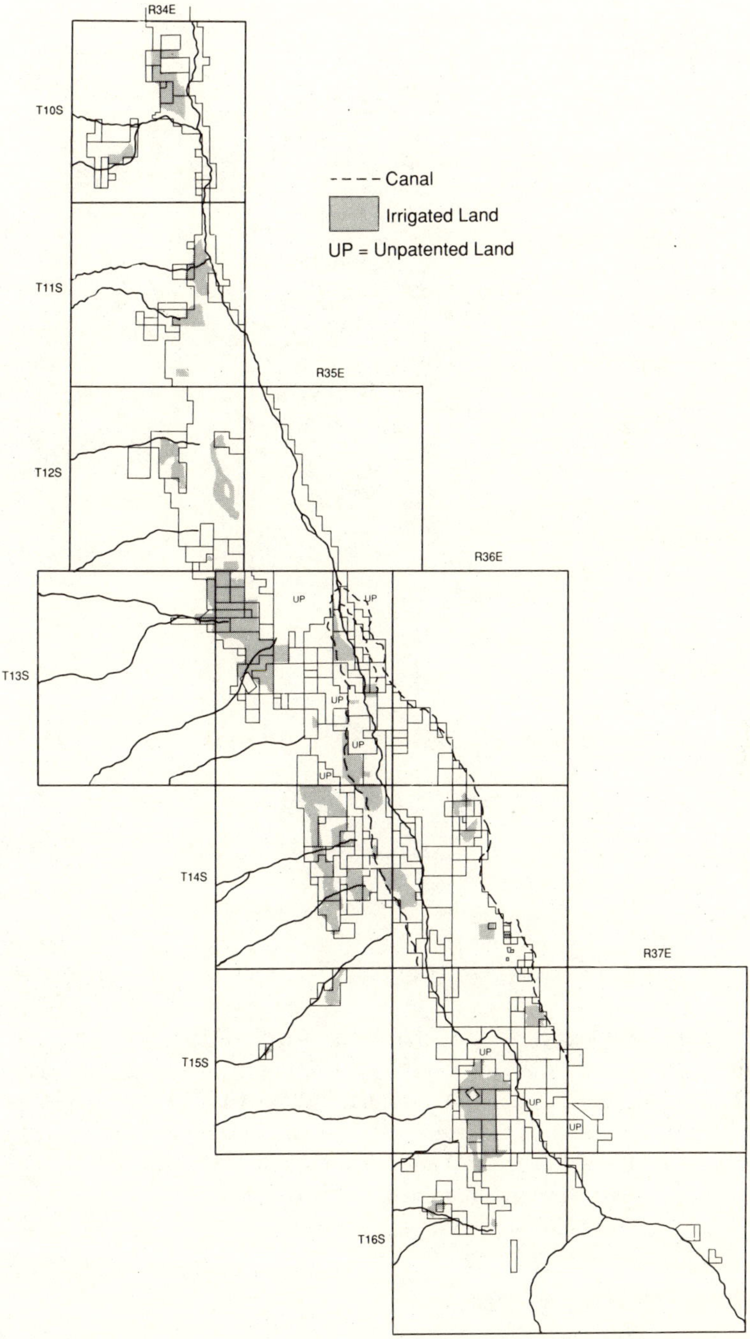
R34E
R35E
R36E
R37E
T10S
T11S
T12S
T13S
T14S
T15S
T16S
Canal
Irrigated Land
UP = Unpatented Land
UP

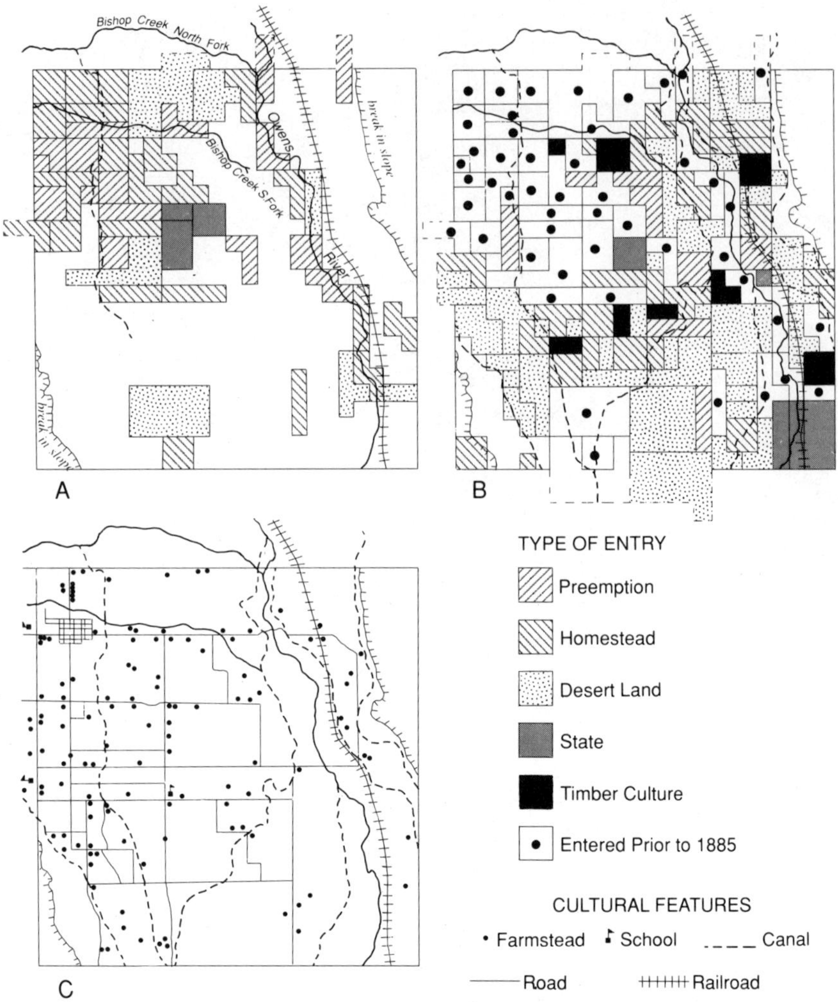

Figure 7.7. Land in T7S R33E: (A) patented land entries through 1884; (B) patented land entries through 1900; (C) pattern of improvements in 1913. (Compiled from Independence District Land Office tract books and Bishop Quadrangle, 1:125,000, U.S. Geological Survey, 1913; reproduced from Sauder, "Powell's Vision of Arid Land Settlement Reexamined," 75, by permission of the Association of Pacific Coast Geographers)

located at the western edge of the valley floor, where water flowing off the Sierra could be easily diverted for irrigation (Fig. 7.8A). By 1900 most of the valley floor in the township, comprising somewhat over half of its area, was under patented entry (Fig. 7.8B). The bulk of the later land selections occurred on the much drier east side of the Owens River, and it was taken in large blocks under the Desert Land Act. Most of the desert land locations were filed in 1887 in connection with the extension of the Inyo Land Company's Eastside Canal. The pattern of settlement and improvement that had evolved in this township by the early 1900s confirms the unsuccessful attempts by the Inyo Land Company to colonize the Eastside tract (Fig. 7.8C). A small cluster of settlers lived in the vicinity of Lone Pine on lands that had been taken early under the Preemption Act, but the sizable acreage patented under the Desert Land Act remained devoid of settlement. The land disposal system was defective in the more southerly township primarily because of the prevailing environmental conditions. In the northern half of the valley, where a relatively dense pattern of early settlement arose from a natural water supply that could be easily diverted for farming, significant numbers of farmers could later join together to construct larger irrigation canals, and settlement could therefore progress. But in the south half of the Owens Valley, where population concentrations were initially restricted by more limited water resources, irrigation developments were much less impressive. The agricultural colonization of the southern valley lagged far behind that of the north largely because of this important environmental difference.

Powell's Recommendations for Arid Land Disposal

Legislators in Congress were unsuccessful in formulating a coherent public land policy for much of the arid West primarily because they failed to recognize the environmental diversity of the region. John Wesley Powell, after many years of Western experience, had become acutely aware of the region's physical attributes and recommended adjustments in the public land laws for disposal of the dry, rugged West.[82] For areas like the Owens Valley's northern townships and isolated locations along the valley's west side, where the settlement of irrigable land depended on the diversion of minor streams by individual effort, Powell believed that the prevailing system of land disposal was adequate. The successful patenting and improvement of these lands in the Owens Valley by bona fide settlers bear this out. But the majority of irrigable land in the Owens Valley's southern townships could not be

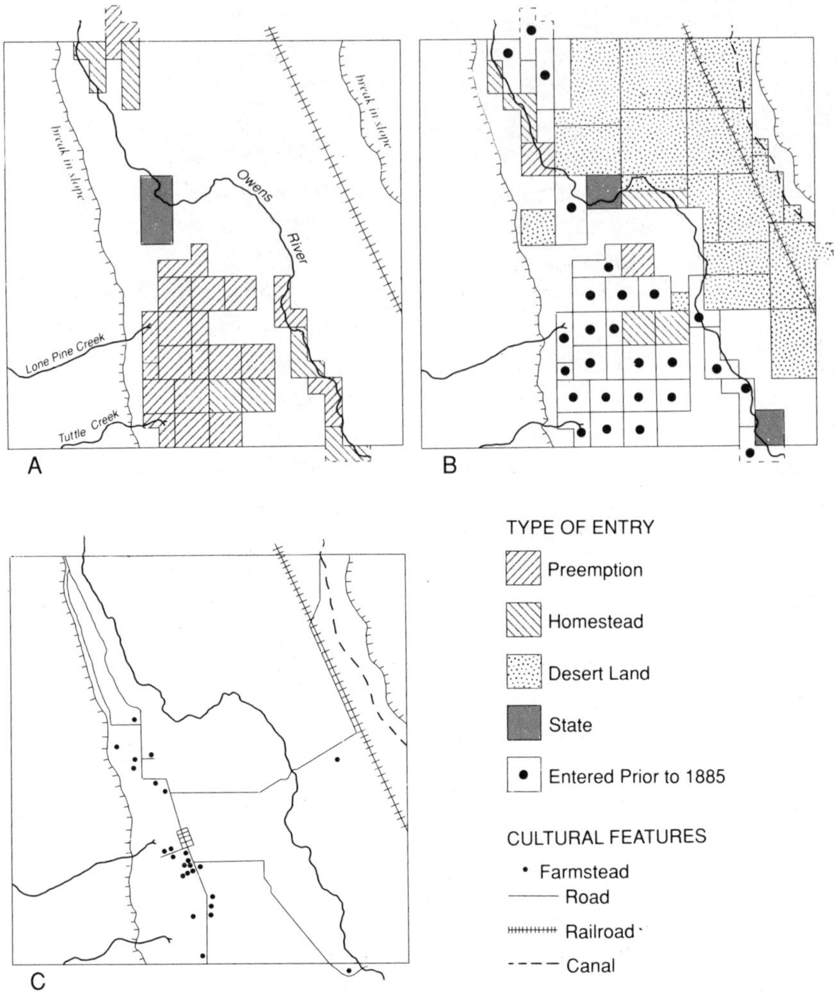

Figure 7.8. Land in T15S R36E: (A) patented land entries through 1884; (B) patented land entries through 1900; (C) pattern of improvements in 1907. (Compiled from Independence District Land Office tract books and Mt. Whitney Quadrangle, 1:125,000, U.S. Geological Survey, 1907; reproduced from Sauder, "Powell's Vision of Arid Land Settlement Reexamined," 81, by permission of the Association of Pacific Coast Geographers)

supplied with water in this fashion. These tracts depended on the diversion of a much larger stream – the Owens River. For poorly watered regions, Powell proposed a cooperative system of land disposal whereby groups of homesteaders would form irrigation districts, establish their own regulations for the use of water and the subdivision of lands within the constraints of the general provisions of the law, and be required to prove that their parcels, which he recommended should not exceed eighty acres, had been redeemed before patent to the land could be obtained. Powell therefore believed that the existing land laws, which promoted settlement by individual rather than cooperative effort, would not succeed in reclaiming the majority of irrigable lands in the arid West.

Powell's observations were accurate when applied to the Owens Valley. By the early twentieth century, individual, corporate and cooperative irrigators in the valley had gone about as far as they could go toward reclaiming the land. The irrigation partnerships that evolved in the Owens Valley built comparatively small ditches that irrigated only a small portion of the valley, leaving the higher alluvial slopes and most of the southern valley unwatered. Had Powell's advice been adopted, the Owens Valley's southern townships might have been more effectively settled by the turn of the century, possibly compelling Los Angeles to look elsewhere to supplement its increasingly inadequate water supply.

8. Federal Reclamation and Municipal Manipulation, 1900–1913

It is a splendid thing to have the great sagebrush forests of Owens valley protected from the ravages of borers and bark beetles.

– "How It Works," *Inyo Independent*, September 3, 1909

Powell's Irrigation Survey

The mining activity at Tonopah and Goldfield brought a new era of prosperity to the Owens Valley, for now the majority of the valley's surplus produce could be shipped to these two booming mining centers. But the region remained isolated from the rest of California because the long-anticipated extension of the Carson and Colorado Railroad to Mojave, which would link the valley to Los Angeles, had not yet materialized. Geographically, the eastern Sierra was more closely tied to Nevada, and the Owens Valley's population, although large enough to have played a significant role in Nevada affairs, had no great influence in the California legislature.[1] According to the *Inyo Register*, Inyo County did not even have the honor of being a "California point" that a cross-country traveler at the time could reach for only $25.00. An additional fare of $18.95 was required at Reno for travel into the valley.[2] The Owens Valley was located in what was politely referred to as "undiscovered California."[3] Few would have thought that it would become the stage upon which one of the nation's most celebrated water disputes was played out.

At the turn of the century, "rivers of water" were literally going to waste in the Owens Valley. In fact, the *Inyo Register* remarked that more water was wasted in the Owens Valley than some of the state's southern counties possessed.[4] The *Inyo Independent* also commented on the water situation, reporting that "An immense volumn [*sic*] of water is running to waste down the Owens River. . . . It seems a pity, with the large area of fertile land still unreclaimed in the valley, that the waste should be permitted to continue."[5] Fred Eaton, who served as mayor of Los Angeles from 1899 to 1901, was acutely aware of the Owens River's surplus flow, for he had lived and worked in the valley prior to assuming the office of mayor. In the fall of 1892 Eaton and his wife arrived in the Owens Valley for an extended stay, hoping that the dry climate would help improve his wife's health. In July of the following year, Eaton purchased a ranch located on Hogback Creek northwest of Lone Pine.[6] By this time the seeds of Eaton's plan to divert the lower Owens River to the growing metropolis of Los Angeles had probably already been planted. In early September 1889, Eaton testified before a Senate special committee in Los Angeles in connection with John Wesley Powell's irrigation survey of the arid lands of the West. In his testimony, Eaton dealt with the waste of water in the Owens Valley and the manner in which water could be more efficiently used, pointing out that in his capacity as a civil engineer, he had been connected with the irrigation works of the valley for some time.[7]

The most revealing testimony pertaining to the Owens Valley, however, came from M. A. Knapp, a civil engineer from Oakland, whose statement immediately followed Eaton's. Knapp presented a plan to the committee for using the surplus water of the Owens River to irrigate the Antelope Valley to the south. In 1888 Knapp had worked as a member of the survey team for the proposed extension of the Carson and Colorado Railroad to Mojave. He was therefore familiar with the region's topography and suggested that "a canal might be run from near Big Pine, on the Owens River, south, past the west side of Owens Lake, passing over a low divide about 9 miles south of the lake, and flowing from there in a natural channel most of the way to Antelope Valley."[8] Knapp explained that, without storage, the area of irrigated land in the Owens Valley was limited to its current level, approximately 30,000 acres, because it was constrained by the water available during the low-water months of April and May and August through October. During the months of June and July, when the melting snows of the Sierra filled the stream channels, not more than one-third of the runoff could be used to irrigate the Owens Valley; and during the winter months the entire flow of the river ran to waste into Owens Lake because large-scale storage was beyond the means of the valley's residents. Even though the water rights in the Owens River were fully appropriated, Knapp suggested that the peak flows of the river would always be too large to be handled by the canals in the valley. It is conceivable that

Eaton's notion of tapping the surplus flow of the Owens River and sending it south via a gravity-flow canal to Los Angeles was inspired by Knapp's testimony, for it was not long afterward that Eaton began to promote his idea of diverting the Owens River to the city.[9] At first Eaton met with little success, but in 1903, when the newly established Reclamation Service began to show an interest in the Owens Valley as a possible reclamation project, the matter became urgent. A Reclamation Service project, if carried forward, would forever put an end to the scheme Eaton had been nurturing for more than a decade.[10]

The Reclamation Service and the Owens Valley

In the Owens Valley, as elsewhere in the West, private enterprise alone was inadequate either to raise the capital required to build large dams to store meltwater runoff or to construct canals and laterals to transport water to higher benchlands. This circumstance led to President Theodore Roosevelt's approval in June 1902 of the Newlands Reclamation Act, which inaugurated a new federal policy on the reclamation of the nation's arid lands. Henceforth the federal government would facilitate the construction of large irrigation projects in the West.

The Reclamation Service's interest in the Owens Valley for a possible irrigation project emerged in June 1903. At that time, J. C. Clausen, a Reclamation Service engineer, was dispatched to conduct a reconnaissance of the Owens Valley to determine the amount and quality of the valley's unpatented public land and to investigate the possibility of storing water for its reclamation.

Clausen's preliminary inspection led to the immediate withdrawal of 21,000 acres to protect a potential reservoir site in Long Valley and an additional 544,480 acres in succeeding months, encompassing nearly all of the remaining public land in the Owens Valley (Fig. 8.1).[11] The purpose of these withdrawals was to prohibit speculators, after learning of the government's interest in an area, from entering claims on land to be served by a proposed project, thereby escalating the cost of the government's land acquisition program should the project ultimately be approved. But Reclamation Service withdrawals did not necessarily prohibit private parties from filing claims on these lands. Two forms of withdrawal existed under the reclamation law. "First form" withdrawals, which included the initial 21,000 acres in Long Valley, were not subject to entry under any circumstances, but "second form" withdrawals, which comprised all of the unappropriated Owens Valley lands, were still subject to entry under the Homestead Act in parcels

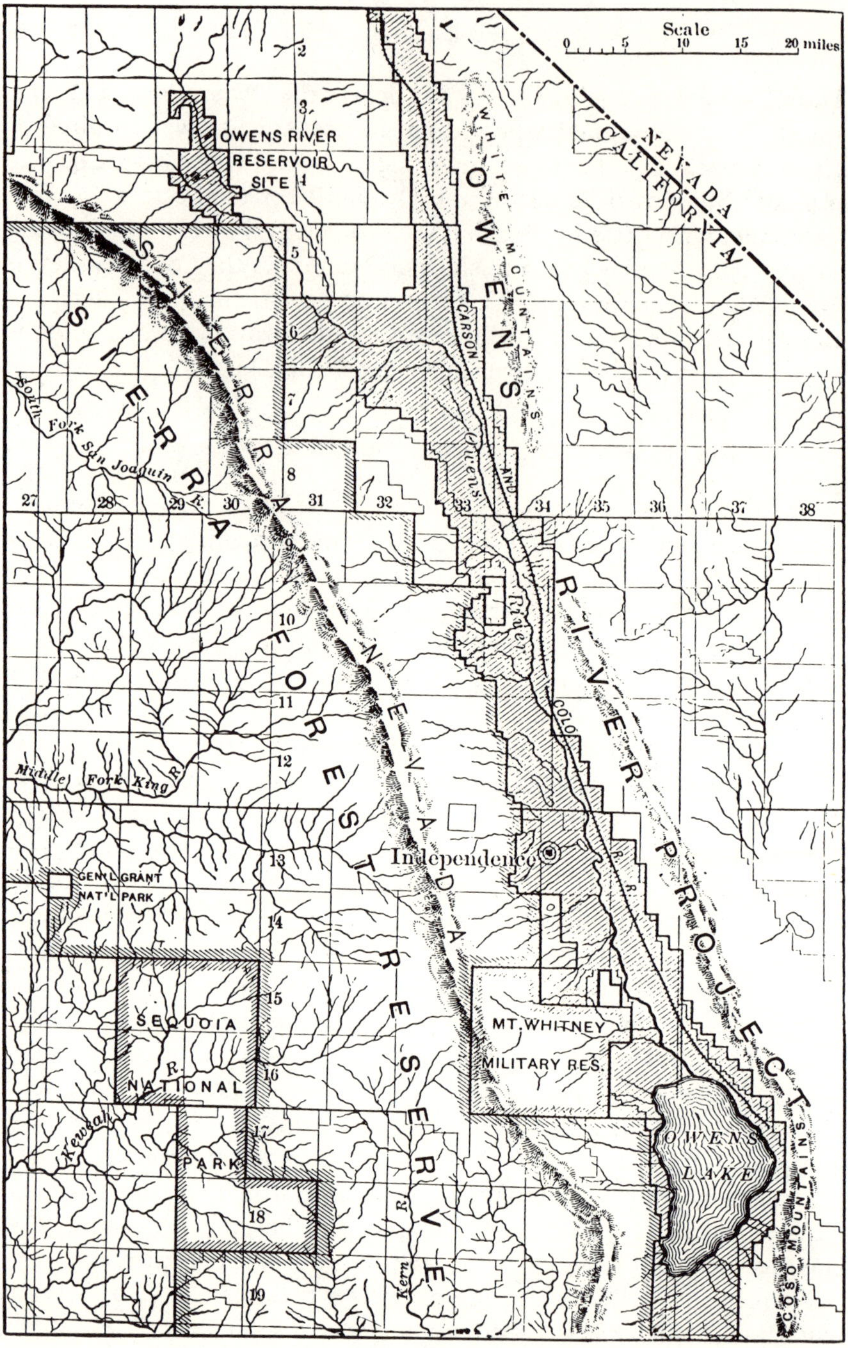

Figure 8.1. Reclamation Service withdrawals in the Owens Valley. (Reproduced from U.S. Geological Survey, *Second Annual Report of the Reclamation Service,* 94)

ranging from 40 to 160 acres.[12] Since most of the valley floor was already blanketed by patented entries amounting to approximately 200,000 acres, the government withdrawals primarily affected higher alluvial slopes – areas that could not be reached by the existing system of irrigation canals and therefore could not feasibly be homesteaded.

Clausen began his measurements of the hydrographic conditions of the valley in August 1903. In November, Joseph B. Lippincott, supervising engineer for Reclamation Service activities in the Pacific Coast region, posted notices on behalf of the government claiming all unappropriated surplus waters of the Owens River.[13] Early in 1904, Lippincott reported that the people of the valley were "extremely anxious to have the government store the surplus flood-waters and regulate the stream flow, not only for the benefit of their lands, the supply of which is now somewhat irregular, but also for the extension of the irrigated areas."[14]

During the following summer, Clausen conducted a detailed survey of the storage possibilities for irrigation in the region, and in November 1904 he reported favorably on the Owens Valley project.[15] Clausen believed that the special advantages for a reclamation project in the Owens Valley included an "abundance of water power, fertile soil, genial climate, nearby markets for all agricultural products in Tonopah and Goldfields and a possible outlet to Los Angeles in the near future."[16] Clausen was convinced of the ability of the valley to produce excellent beef; this alone, he concluded, would warrant federal expenditures to improve conditions in order to double or triple production. But Clausen cited some disadvantages of the valley as well, not the least of which were the thousands of acres damaged in previous years by over-irrigation and the accumulation of alkali. Clausen found that, on average, from 7 to 9 acre-feet of water were applied to farmland during the irrigating season, whereas 4 acre-feet would normally be sufficient. The construction of a drainage system and more efficient use of water would therefore have to be incorporated into any proposed project. Also, if the Reclamation Service was to guarantee an efficient irrigation system, it would have to be given control of the entire water supply of the valley. For this to be accomplished, an association of irrigators would need to be formed to negotiate a plan for relinquishing existing private rights to the federal government. As a show of good faith, more than 400 individuals, who owned a total of 102,433 acres, signed a petition urging the government to proceed as quickly as possible with the irrigation project.[17] Their holdings included about half of the valley's patented acreage.

Thomas Means, the soil engineer for the Owens Valley project, also deemed the project worthy of serious consideration. His survey of the valley resulted in a four-tiered classification system for the valley's soils: first- and second-class soils, totaling 125,000 acres, could be brought under cultivation as soon as water was applied; 80,000 acres of third-class soils would require drainage and washing before cultivation; and fourth-class soils, encompassing 75,000 acres, were not irrigable because of their rough and broken nature.[18] Means suggested that if sufficient water were available, all of the third-class land could be reclaimed, but he recommended that the alkali land located in the Owens River floodplain remain in its current use as pasture. Clausen estimated that with proper storage, drainage, and conservation, enough water would be available during normal years to irrigate a total of 141,000 acres. Since somewhat more than 30,000 acres were then under irrigation, only 106,000 acres of new land could be added, meaning that the majority of third-class lands would remain unirrigated during normal years.

Since the results of the water and soil investigations in the valley proved favorable, Clausen then outlined his irrigation scheme. He recommended that Long Valley be the principal point of storage and proposed the construction of a 140-foot-high dam so that surplus water would always be available to compensate for deficient years. From the reservoir site, water would be channeled through the Owens River gorge to the head of the Owens Valley, where the first diversions would be made into two high-level canals extending along each side of the valley. In order to include the maximum irrigable acreage, each canal would be run as close as possible to the base of its respective mountain range. These upper-valley canals would extend only as far as Poverty Hills and the adjoining lava fields, with the upper west-side canal irrigating 45,000 acres and the upper east-side canal irrigating approximately 25,000 acres. Return flow from upper-valley irrigation would be captured at Poverty Hills and used in the lower valley. Another diversion would be made south of Poverty Hills along the valley's west side to irrigate the remaining 70,000 acres.[19] Clausen estimated that the total cost of the project would be $2.3 million, or $21.58 per acre, based on 106,000 acres of new irrigated land. The cost was significantly below the average of $30.97 per acre for the twenty-eight projects that the Reclamation Service then had under consideration.[20]

Los Angeles Comes to the Owens Valley

Meanwhile, early in 1902, Eaton revealed to William Mulholland, the newly appointed superintendent of the Los Angeles Department of

Water and Power, what he knew about a water supply "adequate for any requirement Los Angeles may ever have."[21] Although initially he was uninterested, late in 1904, while the work of the Reclamation Service was continuing and Los Angeles was suffering from a drought, Mulholland was persuaded by Eaton to travel to the Owens Valley to examine the region's water supply. After his visit to the valley, Mulholland decided to proceed with Eaton's plan as a municipal undertaking, and he began the work of designing the Los Angeles Aqueduct, which would channel by gravity the flow of the lower Owens River 233 miles south to the expanding metropolis. Eaton was informed by Mulholland that Los Angeles would proceed to acquire the Owens Valley water supply and that he should deal directly with the city. There is no evidence that Eaton had as yet acquired any land options in the Owens Valley, and under his agreement with the city he was precluded from speculating in any lands that were needed for the aqueduct. In other words, Eaton reluctantly agreed to sell all the options he obtained for the city for the same price that he had paid for them.

In the spring of 1905, under Mulholland's direction, Eaton went into the south half of the valley to acquire options on key properties bordering the Owens River. His first and most important option encompassed the 45,500-acre Rickey Land and Cattle Company holdings. Approximately half of Rickey's land, amounting to 22,670 acres, was located in the Owens Valley and lay adjacent to the Owens River in townships 10 south through 12 south. The remainder of Rickey's property was in Long Valley, the proposed site of the reclamation project reservoir. Rickey's lands in Long Valley headed Eaton's list of properties for acquisition because, once the city had control of that location, the plans of the Reclamation Service would be seriously impeded.[22] Rickey's Owens Valley lands were also strategically located, for they were properly situated for the aqueduct intake. As a cattle rancher, Rickey had no interest in the proposed reclamation project.[23] He was therefore easily persuaded to sell his extensive holdings.

In April the *Inyo Register* reported that real estate dealings of "large magnitude" were occurring in the south end of the valley.[24] After the contract on the Rickey property was signed in June, the newspaper mistakenly assumed that the purchase was in some way related to the irrigation project and that the project was now "practically guaranteed."[25] It was a common but ill-founded assumption in the valley that Eaton was representing the Reclamation Service.[26] Eaton did nothing to correct the inference that his activity in the valley was related to the government project; indeed, he maintained all along that his acquisitions were made for the purpose of establishing a stock ranch.[27] Eaton kept his real intentions secret, for he knew that if word leaked out regarding the aqueduct scheme, "an army of grafters" would appear, making his job more difficult and the cost to the city much higher.[28] Meanwhile, he pursued his land purchase campaign until he had acquired more than fifty miles of riparian land along the Owens River, extending from Owens Lake almost to Bishop.

On July 29, 1905, news of the aqueduct project broke in Los Angeles, and the *Inyo Register* in its next issue responded with the following headline: "Los Angeles Plots Destruction – Would Take Owens River, Lay Lands Waste, Ruin People, Homes and Communities."[29] The Owens Valley, according to the *Register*, was to be the victim of the greatest water theft on record, converting to wasteland areas that would have become fertile and populous under the reclamation project. The northern valley newspaper understood clearly the potential economic impact on the valley if the proposed diversion were carried out: "When a sufficient proportion of water of any stream is thus sold, the land will be worth little or nothing. If the great majority are either frozen out or sold out and move away, and no possibility for further improvement of the valley exists, what will lands be worth, what will living here become, other than isolation far worse than now?"[30] Eaton justified his activity in the valley by explaining that the majority of the land and water-right purchases lay along that portion of the river where little use was made of the water and that its diversion to other points would cause little change in the present status of agriculture in the valley.[31] Unlike the *Inyo Register*, the *Inyo Independent* seemed little concerned about the Los Angeles water scheme, reporting that the options secured by Eaton included only 3 percent of the valley's cultivated land.[32] Yet Los Angeles's control of the lower Owens River flow would effectively foreclose options on significant future economic developments in the southern half of the valley.

It is a common misconception that the lands acquired in the southern half of the valley in connection with the Los Angeles aqueduct plan were marginal in nature and therefore useless for agriculture.[33] Yet had the reclamation project been completed as proposed, much of this area would have been brought under cultivation. After the initial Reclamation Service investigations of the Owens Valley had been completed, Lippincott stated that "the Valley is large, and contains much high grade public land."[34] According to Clausen's subsequent final report, 70,000 acres located south of Poverty Hills were considered irrigable, and even more land would have been so classified had sufficient water been available for cultivation.

Contrary to popular belief, the southern valley's deficient productivity at the turn of the century had less to do with soil quality than with the failure of the nation's public land system to facilitate settlement there. Although a large body of literature details the methods Los Angeles used to capture the surplus flow of the Owens River, all the accounts overlook the way the public land system expedited the city's land acquisition program, particularly in the valley's poorly watered central and southern townships. Thomas B. Rickey's extensive riparian holdings, entered fraudulently at the state land office by dummy entrymen, was the first large tract to fall to the city. On most of these lieu selections Rickey had paid only a deposit, disregarding the succeeding installments required to complete their purchase. He held no title to these lands; instead he engaged in the practice of holding them while he bought out nearby farms. Shortly after the purchase of Rickey's lands, Los Angeles acquired 12,433 largely unoccupied and unreclaimed acres owned by the William Penn Colonial Association, land that was originally taken by the Inyo Land Company for its Eastside tract under the provisions of the Desert Land Act. Other undeveloped parcels in the southern region, most of which had been patented under the Desert Land Act, also came into the possession of the city. Los Angeles's relatively easy access to the Owens Valley's land and surplus water is largely a legacy of a land alienation process that was poorly adapted to this relatively dry setting. The more favorable environmental conditions found in the north end of the valley produced a relatively dense pattern of small family farms and ranches, a pattern that was resistant, at least initially, to the city's extensive land acquisition program. Had the nation's public land laws achieved greater success in establishing a denser pattern of smallholdings in the valley's southern region, the larger number of irrigated farms and the more intensive use of the Owens River for agriculture might have deterred Los Angeles's advance into the Owens Valley.

Political maneuvering by Los Angeles subsequent to the city's acquisition of strategic valley lands ultimately blocked the Reclamation Service project. The original feasibility of the project was undisputed, but the city's control of fifty miles of riparian land and Eaton's control of the reservoir site made the Reclamation Service's efforts to pursue the project further seem futile. It halted its investigations of the Owens Valley project in January 1906, and in late June President Roosevelt, in the interest of what he called "the greatest good for the greatest number," made the final decision regarding the reclamation project in favor of Los Angeles. A few days later, on June 30, Congress approved a right-of-way bill granting easements to the city through the public lands for the purpose of constructing its aqueduct. Section 4 of the act stated that, with the abandonment of the Owens Valley Reclamation Project, the city would have

> the preference right to acquire at any time within three years from the approval of this Act any lands now reserved by the United States under the terms of said Reclamation Act in connection with said project, necessary for storage or right of way purposes, upon filing with the register and receiver of the land office in the land district where any such lands sought to be acquired are situated on a map showing the lands desired to be acquired, and upon the approval of said map or maps by the Secretary of the Interior, and upon payment of one dollar and twenty-five cents per acre to the receiver of the said land office, title to said lands so reserved and filed on shall vest in said city of Los Angeles.[35]

In other words, abandonment of the reclamation project did not immediately restore to entry any of the withdrawn lands in the Owens Valley. Under the right-of-way bill, the city was given three years to determine which of the valley's remaining public lands would be needed for the aqueduct project. On July 12, 1907, the Reclamation Service officially abandoned the project after Los Angeles paid the agreed upon sum of $14,000 for use of the maps and field notes compiled in the course of the federal survey.

By late 1909, in order to gain an undisputed hold on the Owens Valley's unused surface and underground water supply, Los Angeles had acquired over 82,000 acres of land and appurtenant water rights. This holding encompassed nearly the entire southern half of the valley and also substantial acreage in the northern end (Fig. 8.2). The Rickey lands provided an aqueduct diversion site, but to protect itself against legal injunctions by riparian owners downstream from the aqueduct intake, the city also needed to acquire riparian rights to all lands located south

Figure 8.2. Land acquisitions in the Owens Valley by the City of Los Angeles through 1909. (Compiled from Los Angeles Department of Water and Power, "Map Showing Property Owned in the Owens Valley by the City of Los Angeles," and Los Angeles Department of Water and Power, Aqueduct Investigation Board, "Report," 1912)

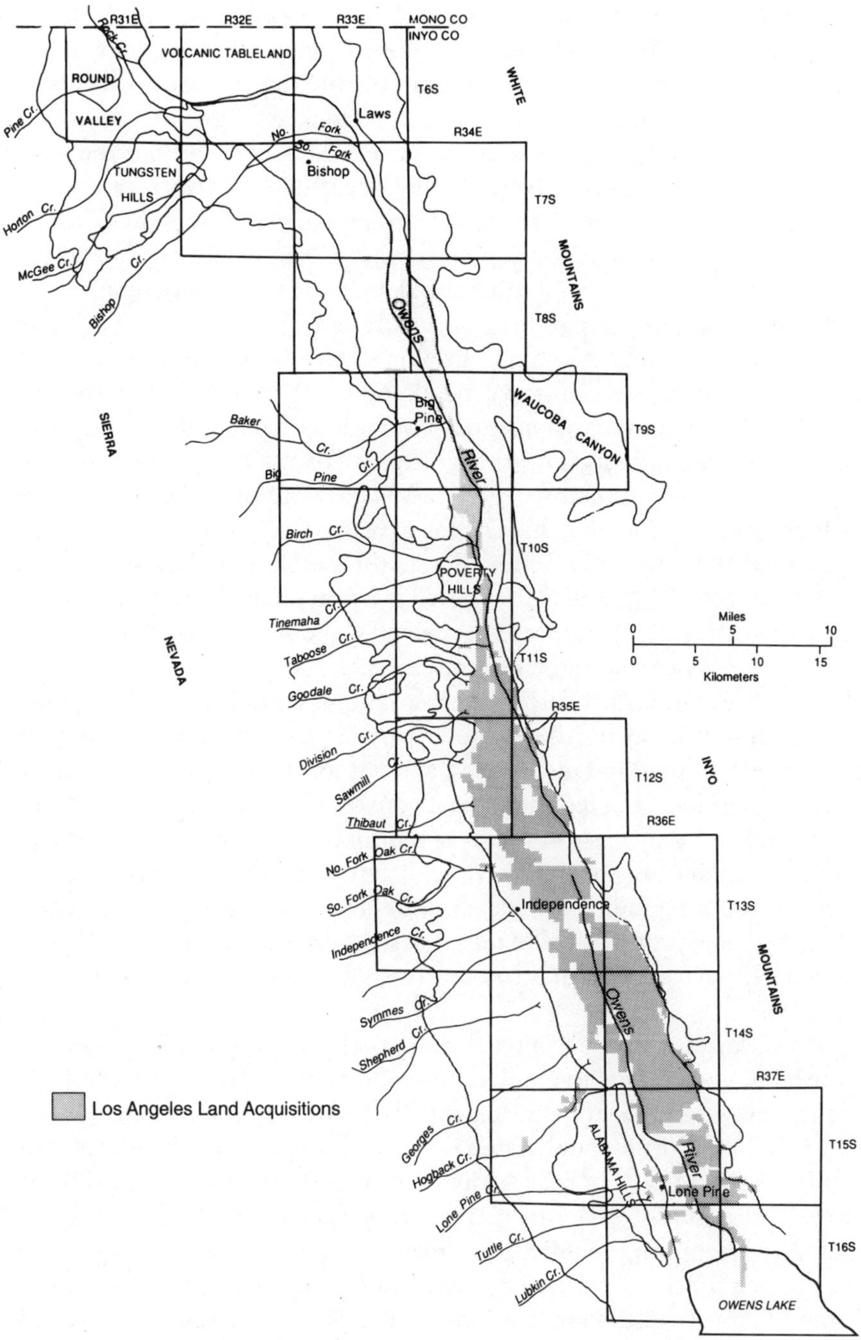
R31E
R32E
R33E
MONO CO
INYO CO
Rock Cr.
VOLCANIC TABLELAND
ROUND
VALLEY
Pine Cr.
Laws
T6S
WHITE
No. Fork
So. Fork
R34E
Bishop
TUNGSTEN
HILLS
Horton Cr.
McGee Cr.
Bishop Cr.
T7S
MOUNTAINS
Owens
T8S
Big Pine
WAUCOBA CANYON
T9S
Baker Cr.
Big Pine Cr.
SIERRA
River
Birch Cr.
T10S
POVERTY HILLS
Tinemaha Cr.
Taboose Cr.
T11S
Goodale Cr.
NEVADA
Miles
0
5
10
0
5
10
15
Kilometers
R35E
Division Cr.
Sawmill Cr.
T12S
INYO
Thibaut Cr.
R36E
No. Fork Oak Cr.
So. Fork Oak Cr.
Independence
T13S
Independence Cr.
MOUNTAINS
Owens
Symmes Cr.
Shepherd Cr.
T14S
R37E
Los Angeles Land Acquisitions
Georges Cr.
ALABAMA HILLS
River
T15S
Hogback Cr.
Lone Pine
Lone Pine Cr.
Tuttle Cr.
T16S
Lubkin Cr.
OWENS LAKE

of the Rickey ranch.[36] By this time California had adopted both the prior appropriation rule (whereby a senior appropriator prevails over a junior one in dry years) and the riparian rule (under which owners of streamside land share water in times of shortage). But Los Angeles's acquisition of extensive riparian frontage did not carry with it the right to export water from the Owens Valley watershed. To divert water from the valley in accordance with the water laws of California, the city was required to file an appropriation notice for the surplus flow of the river.[37]

By mid-1909, about the time the right-of-way bill was to expire, the city's land purchase program in connection with the aqueduct scheme was nearly complete. Although the overwhelming majority of the city's land was concentrated in the central and southern portions of the valley, it had also purchased a nearly contiguous stretch of riparian land extending from T7S R33E to the south line of T9S R34E, amounting to 13,750 acres. Eaton retained 3,400 acres of these lands for himself, on which he would establish his Owens River Poultry Company, as well as water rights to the A. O. Collins Ditch, the Dell Ditch, the Last Chance Ditch, and the Blake and Miller Ditch.[38] Eaton probably was attempting to secure all the river frontage in the valley, but he met with resistance when he reached the more densely settled Bishop region. Ultimately, Los Angeles purchased Eaton's riparian property, and with the procurement of additional northern valley parcels, the city gained a sizable tract north of Big Pine in T8S R33E and T8S R34E (Fig. 8.2). By this time Los Angeles had rejected the idea of constructing a storage reservoir in Long Valley, for to do so would have required the consent of the various ditch companies in the valley, a condition that would have resulted in time-consuming judicial proceedings. Eaton was left holding the Long Valley property, on which he raised cattle, although he did provide the city with an easement for a 100-foot-high dam if one was needed in the future.

Los Angeles instead opted for a lower-level reservoir at Haiwee, south of Owens Lake, to regulate the month-to-month operation of the aqueduct. A second storage site, the Tinemaha Reservoir, would be located in the valley on public land that the city was attempting to obtain under the right-of-way bill. Approval of the Tinemaha land selections were contested by valley residents, who argued that the city was not actually attempting to procure storage facilities but instead was "seeking to acquire land, with the incidental right to extract water therefrom."[39] Los Angeles's applications for the Tinemaha Reservoir site initially were denied by the General Land Office, but upon appeal to the secretary of the interior, the city's request was finally granted.[40] The

Tinemaha Reservoir site was not placed into use until years later, however, after a reservoir at Long Valley was finally constructed. The city's failure to include a reservoir at Long Valley in the early stages of its aqueduct planning was a significant flaw in the design of the project, one that would ultimately trigger the Owens Valley "water war" of the 1920s.

Back in September 1909, the *Inyo Register* had speculated that the reason for the city's purchase of northern valley lands was to obtain as much river frontage as possible in order to establish a claim of riparian ownership, thereby limiting as much as possible the use of water by the irrigation ditch companies.[41] The ownership of riparian lands meant that the needs of the city would be met first before any water would be diverted to nonriparian lands. In an attempt to avert such an occurrence, the *Register* two months earlier had proposed the organization of an irrigation district under the Wright Act of 1887.[42] This would prevent Los Angeles from acquiring ditch stock to the detriment of shareholders who were not in favor of selling. A few months later, when the newspaper renewed its advocacy of an irrigation district, it struck a prescient note. The idea of forming an irrigation district, it declared, "contains one very alluring attraction, as a starter—the water is made appurtenant to the land, and no amount of trading hereafter short of purchase of the whole valley would threaten the valley's development if it were to be put under the district system."[43] Although valley residents had not seriously considered the irrigation district concept for protecting the valley's future at the time, it would reemerge ten years later when the survival of the northern valley's way of life was more imminently threatened by another round of land purchases by Los Angeles.

A Forest Reserve in the Owens Valley

On December 14, 1907—a year and a half prior to the expiration of the right-of-way bill—most of the public lands north of Big Pine in townships 6S R32E, 6S R33E, 7S R32E, 7S R33E, and 8S R33E were restored to entry. Lands in these townships had little impact on the aqueduct project, since they lay far to the north of the proposed right-of-way. The remainder of the Owens Valley lands were still considered "second form" withdrawals of the Reclamation Act and were therefore subject to homestead entry. In order to forestall entry on the withdrawn public lands that might harass the city's project, Los Angeles requested that Chief Forester Gifford Pinchot extend the boundaries of the Sierra Forest Reserve beyond the base of the Sierra Nevada to include large areas of the treeless Owens Valley. This action, approved on February 16, 1907, resulted in the temporary withdrawal from entry of a small portion of the northern valley and most of the valley's southern townships. The precedent for this action had been established on July 25, 1905, four days before the news of the aqueduct project broke, when all of T6S R31E, had come under Forest Service jurisdiction, presumably to protect the strategic Owens River gorge from entry.[44] A common misconception regarding these Forest Service withdrawals, which amounted to approximately 275,000 acres, is that they included areas additional to the Reclamation Service withdrawals in order to prevent further settlement in the valley.[45] This was not the case, at least not in the southern half of the valley. Instead, they overlapped most of the original Reclamation Service withdrawals, merely creating an additional layer of federal bureaucracy that would have to be negotiated by potential land entrants in their attempt to acquire homesteads in the Owens Valley.

Even though laws governing the Forest Service specifically prohibited the inclusion of lands in reserves more valuable for agricultural than forest purposes, the Sierra Forest Reserve extension into the Owens Valley was formally approved by executive order of President Roosevelt on April 20, 1908. Legislation permitting homesteading on forest reserve lands had been passed by Congress on June 11, 1906, but little use was made of this law in the Owens Valley because all applications would henceforth be referred to Los Angeles for review. Protecting the purity of the aqueduct's water supply was one of the alleged reasons for Pinchot's action.[46] Only later did Pinchot admit that the reserve was unfair but that Los Angeles had a need that required protection. In other words, the city's selections had to be made before the reserve could be lifted.[47]

While the Reclamation and Forest Service withdrawals were in force, Los Angeles used its easement for the aqueduct right-of-way to purchase 5,255 acres of federal land at the Long Valley reservoir site, 8,088 acres of channel lands bordering the Owens River in the Owens River gorge, 3,840 acres of vacant public land located in T10S R34E for the Tinemaha Reservoir, and 10,821 acres of artesian lands in proximity to the Tinemaha Reservoir site.[48] With the exception of the easement for the aqueduct right-of-way, the only public land acquired by Los Angeles on the floor of the Owens Valley while the forest reserve blanket remained in force was that connected with the Tinemaha Reservoir. Nevertheless, Owens Valley residents were understandably outraged by Pinchot's action, because the only trees found in the valley, other than the riparian species adjacent to the Owens River, were those they had planted themselves. The *Inyo Register* remarked that tracts lying miles away from the proposed route of the aqueduct were withheld "simply to repress settlement in Owens Valley in order that there may be as little water diverted to valley lands as possible, thus leaving a greater 'surplus' for the city."[49] Meanwhile, the *Inyo Independent*, in an article entitled "Forest Reserves Retard Settlement," lamented that the "rush is passing us by."[50]

The popular notion that settlement and development were repressed in the Owens Valley because of the Sierra Forest Reserve extension is not substantiated by the facts. The majority of land on the valley floor, land that could be reached by the existing network of irrigation canals, had long since been disposed of by the federal government, and the valley's lateral streams were fully appropriated. Consequently, the forest reserve blanket encompassed relatively small areas of previously unpatented public land on the valley floor; in contrast, most of the available public land included in the boundaries of the reserve was located on the alluvial slopes above the existing irrigation ditches. These lands were irrigable, but only by extending higher-level canals such as those proposed by the Reclamation Service along the base of the mountains. This type of system was far beyond the means and ability of valley residents to develop. During the five and a half years prior to the original Reclamation Service withdrawals, from January 1898 to mid-1903, an average of 14 patented entries were filed per year in the Owens Valley, for a total of 78 entries.[51] Seventy-two percent of these entries were located in the better-watered north end of the valley. Little demand for land existed in the valley's southern half, the area most heavily blanketed by the forest reserve withdrawals, and after Los Angeles laid claim to the surplus flow of the Owens River, surface water for irrigation was largely unavailable to prospective farmers in this region. The Owens Valley's economic development had not been placed "in a state of suspension" because of the various withdrawals, as William L. Kahrl suggested.[52] The valley's southern townships had long been in a state of suspension; Los Angeles's appropriation of the surplus flow of the Owens River just sealed their fate.

On February 23, 1911, President William Howard Taft issued an executive order that returned the forest reserve boundary roughly to its original limit at the base of the Sierra. Although the notice of formal restoration of these lands to settlement was not issued until March 1913, the land office began to accept applications for entry on the Owens Valley lands after the initial order. The two-year delay in calling for final restoration was tied to action still pending on power applications in the Owens River gorge.[53] Once the city received final patent to these lands in January 1913, the forest reserve blanket was lifted. Reopening the Owens Valley lands to settlement produced an onslaught of promotional literature touting the prospects for the valley's economic development.[54] Nevertheless, relatively few patented entries were filed in the Independence Land Office after the land entry freeze was lifted, and many of these were speculative in nature, for they were made in T16S R36E in proximity to the Southern Pacific's standard-gauge branch railroad, completed in October 1910 to facilitate aqueduct construction. The new line spanned the void between Mojave and the narrow-gauge Carson and Colorado, now known as the Nevada and California Railroad. The Owens Valley finally had its rail connection with Los Angeles.

Beneficial Use

When surplus water from the Owens River spilled into the Los Angeles Aqueduct in November 1913, its flow exceeded the contemporary domestic needs of Los Angeles by about four times.[55] However, in accordance with California water law, Los Angeles was required to put the entire aqueduct flow to beneficial use to avoid endangering its appropriative claim, the basis upon which it was exporting water from the valley.[56] Since Los Angeles acquired the surplus flow of the Owens River in anticipation of continued growth, the aqueduct flow in excess of immediate domestic requirements was turned to irrigation, not in the Owens Valley but in the soon-to-be-annexed San Fernando Valley. In January 1906, Owens Valley residents attempted to prevent such use of their water. At that time their congressional representative introduced a compromise proposal that would let the Reclamation Service proceed with its project, allowing Owens Valley farmers to receive their share first, with the remainder going to the city for domestic purposes only. After both needs were satisfied, any additional water would be used for irrigation in the Owens Valley. Owens Valley farmers had acquiesced to municipal use of their water, but they adamantly opposed its use for irrigation elsewhere.

Acceptance of this proposal, however, would have meant that as Los Angeles's need for water grew with its population, in William Kahrl's words, "it would have to fight the valley in court for every additional drop it took from one year to the next."[57] Los Angeles's undertaking was based on the understanding that the entire aqueduct flow would not be used immediately for domestic needs but that consumption by the growing city would increase over time. In spite of the city's explosive growth—reflected in a tripling of its population from 102,479 in 1900 to 319,198 in 1910—Los Angeles had neither the intention nor the ability to use its aqueduct water solely for domestic purposes. One reason why Los Angeles did not pursue the more conventional course of condemnation proceedings in the Owens Valley but instead moved quickly and quietly to purchase riparian holdings involves the city's inability to show required proof that it absolutely needed the water solely for domestic purposes.[58]

In an attempt to resolve the issue, the secretary of the interior commissioned a study to determine whether the Los Angeles aqueduct plan and the federal project could coexist. The conclusion was that they could not.[59] Acknowledging the probability that the reclamation project would be abandoned in favor of Los Angeles, valley residents put forth another proposal that involved a more modest reclamation effort using Fish Slough at the northern edge of the valley as a potential reservoir site.[60] This was a significantly less expensive proposition for the Reclamation Service but was still beyond the means of valley residents to undertake as a private enterprise. This proposal was also rejected, and the subsequent use of the Owens Valley's water for irrigation in the San Fernando Valley while the valley's southern townships remained barren would forever polarize city and valley interests. President Roosevelt regretted the despair and disappointment that prevailed among Owens Valley residents, but he believed that more valuable use of the nation's resources required that the valley be sacrificed to what he called the "infinitely greater interest" of Los Angeles.[61]

Pondering the Reclamation Service Project

It is curious to ponder the outcome of the Owens Valley Reclamation Project had Los Angeles not intervened. Only three of eight projects under investigation in California at the time were eventually authorized: the Yuma project on the California–Arizona border, the Klamath project on the California–Oregon border, and the Orland project in the Sacramento Valley. Under the original terms of the Reclamation Act, projects were to be supported by funds derived from the sale of public lands. That is, 51 percent of the proceeds from public land sales in each Western state would be dedicated to reclamation projects located within that state. Once the projects were placed in operation, the beneficiaries were expected to reimburse the government for the costs of construction over a ten-year period. When the Owens Valley project was formally abandoned, all available funds for California projects were temporarily exhausted because the Yuma and Klamath projects, approved in 1904 and 1905 respectively, were now being developed.[62]

Notwithstanding the temporary funding problems, the Owens Valley project probably would have received final approval because the natural conditions and patterns of development in the Owens Valley were not unlike those found in the Klamath River region. The Klamath region, like the Owens Valley, was already partially settled prior to the Reclamation Service showing an interest in it, and the first step there, as was recommended for the Owens Valley, was to harmonize the vested rights in land and water, giving the federal government control over the water supply.[63] Both the Owens Valley and Klamath regions were situated at an elevation of 4,000 feet, and the products of each were restricted to grasses and grains, as well as some vegetables and hardy fruits. Alfalfa was the leading crop of both areas, generally yielding two cuttings annually. The production of alfalfa, in combination with plenty of free range surrounding each district, assured the permanence of an important livestock industry. Ostensibly, the underlying reason for the approval and construction of most federal reclamation projects, other than cost, was the prospect they held for development of the livestock industry, because today the most common crop grown on federally watered farms is forage to feed cattle.[64]

When all the variables are weighed, including the enthusiastic support by valley residents for a federally sponsored reclamation project as well as its comparatively low estimated cost, final approval of the Owens Valley project seems almost certain. Had the reclamation of the valley been carried forward as proposed, the region's elevation, 152-day growing season, and limited expanse would have hindered its development into one of California's leading agricultural districts, but the project most certainly would have brought a new life and prosperity to the Owens Valley, particularly to its southern townships. Instead, agriculture henceforth would largely be restricted to the northern end of the valley. But here, too, it was destined to retreat as Los Angeles gradually tightened its grip on the valley's water supply.

9. The Intensification of Agriculture, 1900–1925

The trend of the times . . . is toward the cutting up of large ranches into smaller farms . . . and . . . farming on a large scale, such as is conducted on cattle and grain ranches, will gradually give way . . . to the more intensified methods of the smaller farmer.

– "In the Land of Plenty," *Inyo Magazine,* August 15, 1908

Inyo County is just awakening to its limitless possibilities.

– California Development Board, 1917

Smaller Farms and Increased Specialization

After the aqueduct's completion in 1913, Owens Valley residents settled in for a period of uneasy coexistence with Los Angeles.[1] The diversion of the Owens River's surplus flow to the city affected the existing status of agriculture in the valley very little, since most farming was located north of the diversion site. But future agricultural developments were limited predominantly to the northern end, where appropriative rights senior to those of Los Angeles prevailed. Since the city had tapped the river below the valley's primary center of agriculture, under ordinary circumstances both farmers and city dwellers would receive their required allotment of water. No longer, however, would developments in the valley proceed at their own pace. Los Angeles now played a significant role in the region's economic evolution.

At the time that the water conflict was simmering during the early 1900s, the Owens Valley found itself in the midst of unprecedented prosperity. The rise of Tonopah, Goldfield, and other smaller mining centers in southwest Nevada and the new railroad link to southern

Table 9.1. Number of Farms, Farm Acreage, and Improved Acreage in Inyo County, 1900–1925

	1900	1910	1920	1925
Number of farms	424	438	521	482
Average Farm Size (acres)	333	252	269	299
Acres in Farms	141,059	110,142	140,029	144,182
Improved Acres	43,740[a]	38,698	39,904	n a

SOURCES: U.S., *Twelfth Census, 1900, Agriculture,* pt. 1; U.S., *Thirteenth Census, 1910; Agriculture 1909 and 1910,* vol. 6; U.S., *Fourteenth Census, 1920, Agriculture,* vol. 6, pt. 3; U.S., *Census of Agriculture, 1925,* pt. 3.

[a] This figure is approximately 2,700 acres greater than the irrigated acreage reported in the Twelfth Census, 1900.

California provided markets for all the produce valley farmers were able to supply. In response, Inyo County's population increased 60 percent to nearly 7,000 during the first decade of the twentieth century. Since the majority of the lower valley was controlled by the city and the tracts that could be served by the extensive network of irrigation canals in the Bishop region were now under patent, new arrivals who intended to farm in the Owens Valley were forced to purchase smaller tracts from established landowners. Consequently, one of the more significant developments in the early twentieth century involved an acceleration of the process of farm subdivision. The average farm size in the Owens Valley decreased from 333 acres in 1900 to 252 in 1910 (Table 9.1), but even more illustrative of the move toward smaller farms is the percentage of farms in the valley encompassing fewer than 100 acres: 43 percent in 1910 compared to 31 percent in 1900 (Table 9.2).

The Owens Valley's newspapers considered the move away from general farming to more intensive, specialized farming to be a positive sign for the valley's future and encouraged the process. In 1907 the *Inyo Independent,* for example, reported that the annual profit on one valley farm that produced a bountiful crop of apples, potatoes, corn, wheat, barley, alfalfa, and honey was "a small part of what it could be made to yield by intensified farming."[2] A theme that wove through the majority of local newspaper articles dealing with farming in the region was that the Owens Valley must not only shift to more intensive farming but must also begin to produce more valuable crops. "By means of intensive farming and a close study of market conditions and of the products adapted to the soil and climate," advised the *Independent,* "the value of the product per acre can be greatly increased."[3]

The intensification of agriculture in the Owens Valley was a natural response to the development of irrigation in the northern townships, but the process was also enhanced by the completion in 1910 of a standard-gauge rail line linking the Owens Valley with Los Angeles. A number of colonization schemes followed in the wake of the new line, more than the valley had seen in recent decades. Surprisingly, one of the most ambitious colonization efforts was located in the southern half of the valley in an area untouched by Los Angeles's acquisitions. Taking advantage of the waters of Shepherd Creek flowing off the Sierra five miles southwest of Independence, a syndicate calling itself the Owens Valley Improvement Company began to subdivide the old Shepherd Ranch into farms of from sixteen to twenty-five acres and platted the new town of Manzanar near the center of the subdivided tract. Although Manzanar failed to develop as planned, considerable demand existed for the small tracts surrounding it, most of which were marketed in southern California. The company's promotional literature stressed the importance of the recently completed rail connection with Los Angeles, which, it said, "opens up for the first time this marvelously fertile val-

Table 9.2. Number of Farms by Farm Size in Inyo County, 1900–1925

	1900		1910		1920		1925	
	No.	%	No.	%	No.	%	No.	%
Less than 50 acres	90	21.2	113	25.7	153	29.4	192	40.6
50–99 acres	40	9.4	75	17.1	86	16.5	71	15.0
100–499 acres	248	58.5	208	47.5	236	45.3	158	33.4
500–999 acres	28	6.6	22	5.0	27	5.2	29	6.1
1,000 acres or more	18	4.2	20	4.6	19	3.6	23	4.9[a]

SOURCES: U.S., *Twelfth Census, 1900, Agriculture,* pt. 1; U.S., *Thirteenth Census, 1910, Agriculture 1909 and 1910,* vol. 6; U.S., *Fourteenth Census, 1920, Agriculture,* vol. 6, pt. 3; U.S., *Census of Agriculture, 1925,* pt. 3.

[a]Although 482 farms were counted in the 1925 Census of Agriculture, only 473 farms were reported by farm size.

ley for immediate settlement and culture, and puts the Owens Valley farmer in quick touch with the best markets."[4] The Owens Valley Improvement Company's general plan was to expand the valley's apple production, and the firm therefore offered to plant apple trees and care for them for absentee landowners or to sell the trees directly to residents (Fig. 9.1).[5] It was here also that the Owens Valley's first coordinated attempt at water conservation was put into effect. In contrast to the methods of irrigation that were employed elsewhere in the valley, to prevent seepage and improve drainage the new irrigation system at Manzanar consisted of miles of concrete and tile pipe.

No other significant new districts of compact settlement were created in the valley. More commonly, existing farms in the more populous Bishop region were subdivided into smaller parcels, where land with water rights was selling for prices ranging from $75 to $150 per acre.[6]

Dairy Farming

Attempts to intensify farming and increase specialization in the Owens Valley revolved around three primary activities—dairying, sugar beet production, and fruit growing. The rapid advance of dairying in the valley's north end was reflected in the numerous concrete silos that began to appear in the Bishop region around 1915. Roberts's silo in Round Valley, constructed in 1898, remained for a long time the only one in the valley, but between 1914 and 1916 about a dozen silos were installed, and "many more were in the course of construction."[7] By 1922 more than a hundred silos were dispersed across the valley floor, most in the northern end of the valley.[8] The practice of making ensilage caught on in the Owens Valley partly because it reportedly increased milk production by 30 percent. But in spite of the adoption of corn ensilage and the abundance of hay produced in the valley, dairying remained a small-scale enterprise here compared to other parts of the state. A significant drawback to dairying was the quality of the dairy herds. A dairy herd in the Owens Valley, for example, produced only 53 percent as much milk as one of comparable size in Humboldt County, largely because many of the valley's farmers were trying to raise what was jokingly referred to as an "all round cow" by crossing beef with dairy cattle.[9] Dairy farming, like so many other enterprises in the Owens Valley, was hampered because of the valley's habitual isolation, which also tended to foster unprogressive attitudes among many old-time farmers.

Figure 9.1. Fruit orchards at Manzanar in 1931. Following the abandonment of Manzanar, the subdivision in 1942 became the site of a World War II relocation center housing 10,000 Japanese-Americans. (Courtesy of the UCLA Department of Geography Aerial Photo Archives)

Beet Culture

When the process of farm subdivision was getting underway in the valley, the sugar beet industry was considered as a possible new form of specialized production. Discussion of the prospect of locating a beet sugar factory in the Owens Valley began in 1916 with the Holly Sugar Company, but before any decision could be made, experiments with beet culture had to be undertaken. The *Inyo Independent* suggested that if the thousands of acres currently devoted to grazing were converted to beet production, the return per acre would increase dramatically.[10] Farmers were encouraged to plant sugar beets for the next two years in order to test the feasibility of beet production in the valley. The Holly Sugar Company furnished the seed and hoped to see 1,000 acres planted to beets in 1917 so that at the time of harvest 500 tons of beets a day could be shipped to its plant in Huntington Beach.[11] In the fall of 1916, a field superintendent for the company traversed the valley, signing up acreage for the following year's crop and instructing farmers in the methods for raising beets profitably.[12] Farmers were told that if a beet factory were to be attracted to the Owens Valley, at least 10,000 acres would eventually have to be devoted to beet production. This amounted to somewhat less than the total acreage planted to alfalfa, at the time the valley's leading crop. The hot summer days and cool nights of the valley were ideal for plants to yield a high percentage of sugar, but beet production was significantly more labor-intensive than alfalfa production. Obtaining sufficient labor to handle the crop was therefore an obstacle to its adoption. Consequently, in 1917 only 400 to 500 acres of beets were planted,[13] and although by the end of the year beet culture was proclaimed a success, the valley's total yield of 2,500 tons was not sufficient to establish a sustained interest in the crop. Beet culture therefore quickly faded from the Owens Valley scene. In 1920 only 13 acres were planted to sugar beets, yielding just 33 tons.[14]

Fruit Growing

With the failure of the beet sugar industry and the inherent obstacles to the expansion of dairying in a beef-producing region, the best hope for the future development of the valley was its expanding fruit industry. As the decade of the 1920s approached, a widespread feeling emerged that it was finally time for Owens Valley farmers to focus on the production of one type of crop, a crop that would allow the valley to establish a reputation for itself. If Fresno County was noted for its grapes, and southern California for its citrus, how could Inyo County excel? Apple growing might have been a logical choice. In 1917 it was regarded as the most promising and important crop in the valley after alfalfa, and Owens Valley apples had won numerous medals at various fairs and exhibits over the years.[15] The Owens Valley's alluvial slopes and climate seemed to be well suited to apple production, but Owen's Valley's emphasis on this activity would not be unique in the West. Apple production was by now firmly established in Washington state, and closer to home Victor Valley, just 100 miles from Los Angeles, had already emerged as an important apple producing region.

Although the acreage devoted to apple production did expand significantly during the early 1920s, particularly at Manzanar, many in the valley believed that pears held greater promise. The California Development Board concurred. In 1917 the Inyo County Board of Supervisors had commissioned the Development Board to compile a study of the county that would provide unbiased "facts of interest and guidance to the prospective home seeker and settler."[16] After surveying the Owens Valley, the board concluded that pears would eventually become the leading fruit crop of the county.[17] In 1920, at the same time that Los Angeles was beginning a new campaign to divert additional water from the Owens Valley, a crusade to encourage the expansion of pear production swept through the northern end of the valley. It was now more important than ever, cautioned the *Inyo Independent,* that the Owens Valley concentrate on the production of a high-value specialty crop. Otherwise the water that was currently appropriated by valley farmers would flow to areas where it would be used more profitably.[18] High-priced land in the valley, the newspaper argued, would "absolutely" make water rights safe from encroachment by outside interests, and it suggested that the valley's salvation would be pear culture.[19]

At the time, pears were considered the most profitable of the deciduous fruits, because the established pear orchards of the Midwest and Northeast had been severely damaged by the pear blight, and pear production was now shifting westward into California. To avoid blight, the U.S. Department of Agriculture recommended planting commercial pear orchards at elevations ranging from 3,000 to 5,000 feet. Although the pear tree grew more slowly at this elevation, it would be a more hardy, sturdy tree, resistant to the attacks of the blight. The Owens Valley's special advantage for pear production lay in its elevation, and although pear culture had previously been attempted on only a small scale, pears had never failed in the valley. Valley farmers were therefore encouraged to seize this historic opportunity and leap to the forefront of this developing enterprise. No one suggested that the Owens Valley be devoted exclusively to pear culture; instead, pear production would be used as a means of increasing the value of "every foot of land" within the valley, thereby putting its economy on a firmer foundation. This enterprise would give greater security to everyone's investment, it was argued, because pears would provide sufficient returns to keep the water in the valley.

To stimulate interest in pear culture, the Inyo County Pear Growers' Association was organized. Members were required to subscribe fifty cents an acre per year for two years to further the enterprise. Valley residents must not have embraced the idea immediately, however, for subscriptions accrued slowly. In December 1920 the *Owens Valley Herald,* published in Big Pine, made the following plea:

> The future of the valley depends upon the success of this organization. It is the first organization for the advancement of the valley ever started . . . [and] unless a strenuous effort is made in Owens Valley to make the land here productive enough in crops that pay big returns, then the water will be taken to some other place where it will do this.
>
> If they [the landowners] do not grasp the present opportunity as it has presented itself, this country will be killed in order to give added life to some other place that really wants to live and grow.
>
> Have enough interest in your own welfare and the welfare of the valley to do your part, and do it now.[20]

This admonishment apparently achieved the intended results. During the spring of 1921 nearly 20,000 pear trees were planted in the valley.[21] The Owens Valley was undergoing "a kind of evolution," reported the *Owens Valley Herald,* "which must always be undergone by a country when it begins to awaken to its enormous possibilities."[22]

The *Los Angeles Times,* in its annual midwinter edition in January 1922, confirmed the new direction that agriculture was taking in the valley when it published the following:

> Increased interest in fruit culture, as shown by the unusually heavy planting of apple and pear trees last season and the promise of equally heavy planting the coming season, indicates that Inyo County will soon become one of the important fruit centers of the State. Soil and climatic conditions are especially adapted to the growing of red apples and Bartlett pears, giving them keeping qualities and flavor that are unexcelled. These fruits have been grown in the Owens Valley for many years and the industry is now being put on a sound commercial basis through the efforts of the Inyo County Farm Bureau and the Owens Valley Pear Growers Association.[23]

By now, orchard planting in the valley was progressing at an almost feverish pace. The number of fruit trees planted during the winter of 1922 was nearly double that of the previous year, with the most striking growth reflected in pear plantings.[24] Between 1920 and 1925 the number of bearing pear trees in the valley multiplied elevenfold, from 3,405 to 37,753, while the number of bearing apple trees tripled during the same period, increasing from 17,963 to nearly 52,621.[25] Based on an average of 70 plantings per acre, the estimated number of acres devoted to pears totalled 540, while apple orchards occupied nearly 900 acres. With the exception of Manzanar, many of these emerging orchards clustered on the higher, relatively frost-free fan deposits in the West Bishop and Sunland districts adjacent to the Owens River Canal, where subdivided parcels were selling for as much as $350 per acre in 1920.[26] With the widespread adoption of horticulture, a general optimism permeated the entire valley: "Signs point to a greater activity during the summer months in the way of ranch improvements than has ever been seen here before, and altogether the future looks exceedingly encouraging," reported the *Inyo Independent* in the winter of 1922.[27]

Recurrent Problems

In spite of the prevailing optimism, two age-old problems continued to plague the region: transportation and drainage. By the mid-1920s, Los Angeles had become the principal market for surplus products of the valley, but freight charges remained high because the old Carson and Colorado tracks, which extended the length of the valley, had not been converted to broad gauge by the Southern Pacific when it completed the trunk line to Mojave. Freight from the Bishop region therefore still had to be transferred from narrow- to standard-gauge cars at the junction north of Owens Lake, adding to the cost of the journey. Although the rail link with Los Angeles unquestionably influenced agricultural developments in the valley after its completion, a more efficient standard-gauge line traversing the entire length of the Owens Valley never materialized.

Drainage, however, was an even more serious impediment to the region's development. In 1917 the *Inyo Register* reported that "it is accepted as a fact that the water table of the valley is higher than it used to be; that some of our lands once sage-covered have become almost wholly saturated. This is a condition that has grown gradually, and that will continue to grow worse — a little more each year to the great detriment of many good properties."[28] By the early 1920s approximately 21,000 acres, or 60 percent of the land in the Bishop region, required drainage.[29] The water table throughout most of this area averaged six feet or less, but water was within two feet of the surface over approximately half of the region. The Owens Valley's naturally poor drainage, resulting from its impervious nature, had been aggravated by years of over-irrigation from unlined canals constructed across porous soils. Further, because the valley lacked any storage facility, water for domestic and livestock needs required a continuous flow throughout the year, preventing the water table from receding during the winter months, as was common in most irrigated regions.[30] But most important, the farmers of the valley had not altered their habit of profligate irrigation. An abnormal supply of water for irrigation continued to be used on fields that had never undergone preliminary grading or leveling. The California Development Board summarized the latter problem in 1917 when it bluntly noted that "the duty of water is something of which the average farmer in this section knows little – or cares less. Few attempts are made at water measurement or gauging the amount necessary for various soils and crops."[31] For example, the efficiency of agricultural water use among the farmers diverting water from Bishop Creek was estimated at the time to be just 20 percent.[32] Consequently, less than one-quarter of the water applied to the soil sustained crop growth, while the remainder was lost to seepage and evaporation.

Although some farmers in the valley had attempted to undertake small-scale drainage, their efforts were hampered by the lack of satisfactory outlets for the discharge. Only a large, coordinated system of drainage involving the entire community of farmers would bring about a general lowering of the water table. In an effort to encourage such coordination, the California Development Board proposed that a drainage system be installed in the upper end of the valley, encompassing approximately 20,000 acres and 200 to 250 farms. At the board's request, the U.S. Department of Agriculture sent two engineers to the Owens Valley to investigate the feasibility of such a system. Their survey confirmed that drainage should be undertaken; in fact, it revealed that the Owens Valley was in greater need of drainage than any other region that they had studied.[33] The estimated cost for farmers living in the proposed drainage district averaged twenty dollars per acre, but if the district issued bonds, the cost could be paid back over a twenty-year period. Bearing such a burden was difficult for valley farmers to accept, for the cost of originally diverting the water was small in comparison. Discussion of the drainage issue dragged on for months. Finally, early in 1919, the *Inyo Register* reported that there was "no special encouragement for hopes of adoption of a general [drainage] system," and the issue was dropped.[34]

The Department of Agriculture engineers had warned that persuading farmers to organize would prove to be the most difficult obstacle to the formation of a drainage district but that "to accomplish results, it will be absolutely necessary for the people to co-operate."[35] Because of the valley's isolation through the years, it had developed into a fortress of rural conservatism. The pioneer sense of individualism that had been spawned at the time of settlement and that had become firmly established in the valley worked against such large-scale cooperative endeavors. Granted, the upper valley's irrigation canals were monuments to the cooperative efforts of its early settlers, but the farmers constructed the canals with their own labor, keeping the canal enterprises free from indebtedness. Further, cooperative action was comparatively easy to arrange for canal construction since all the cultivators required a water supply. In contrast, cooperation was significantly more difficult to achieve in connection with drainage because not all the irrigated land became waterlogged simultaneously. The process was a gradual one in which the lower-lying farms suffered first. Since the irrigators in a district did not have equivalent and concurrent needs for drainage, many did not understand or appreciate the value of a coordinated system. The formation of a drainage district was therefore a much more ambitious and provident endeavor, one which was too farsighted to gain a consensus among a majority of Owens Valley farmers.

In spite of the drainage setback, momentum continued to build for what was hoped would become one of the premier fruit growing districts in California. By the mid-1920s the northern valley's acreage under cultivation had not changed significantly from previous years, but the widespread adoption of horticulture had dramatically altered the valley's cultural landscape. The intensification of agriculture in the northern end of the Owens Valley resulted in a more compact pattern of settlement, almost as if to insulate the region from the fresh advances of the urban giant to the south. But the move toward intensified farming and crop specialization to improve land values and thereby safeguard the farmers' water supply proved unsuccessful. Los Angeles had gained a foothold in the valley years earlier, and it was only a matter of time before it would once again come knocking on the Owens Valley's door to augment its ever more inadequate water supply.

10. The Lost Frontier

> The economy and town life were suspended on a wire of doubt: investment was withheld, lending stopped, real estate trading stagnated save for the city's bidding, farm production dropped precipitously, schools began to close.
>
> —John Walton, "Picnic at Alabama Gates"

> We, the Indians of Independence Fort, Calif., Inyo County, say once we Indians had our ways and had own food and meat and we owned this valley—Owens Valley. When the white people come they took away everything from us and learned us how to work and we worked for them. Now they took the work from us, we are without work now. We want work to be given to us, we the Owens Valley Indians, right in Owens Valley.
>
> —John Symmes, *Survey of Conditions of the Indians*

Drought Conditions and the Need for Storage

The year 1922 dawned on an optimistic note in the Owens Valley. "From an agricultural and horticultural standpoint," commented the *Inyo Independent* in January, "the future of Owens Valley looks particularly bright."[1] Three years later to the day, however, the following headline appeared in valley newspapers: "Greed of City Ruins the Owens Valley."[2] The valley had gone from the pinnacle of hope to the depths of despair and devastation in the short span of three years.

A prolonged drought plagued the entire Southwest during the early 1920s. Runoff from the Sierra was far below normal, and after Owens Valley farmers had made their diversions for irrigation, there was little water left to supply the Los Angeles Aqueduct. Los Angeles's population was now approaching the one million mark, and coincident with the city's phenomenal expansion came an ever-increasing demand for water. Without the Long Valley reservoir there was no means of storing the precipitation of wet years; and when the dry years came, there was insufficient water to supply both the valley and the city. During the summer of 1923, when the aqueduct flow was about half its normal level, the drought situation reached crisis proportions in Los Angeles.

As early as 1918 the city had begun to take steps to protect the aqueduct supply against a possible time of shortage by sinking wells on its property in the vicinity of Independence. In lieu of constructing a storage reservoir in Long Valley, the city opted to tap the valley's natural underground reservoir. By 1920 these underground withdrawals had ignited a new conflict in the valley as farmers living in the Independence region began to notice damage to their land resulting from the city's pumping. Once the farmers discovered that their underground water supply was being siphoned off by pumps on adjacent property owned by the city, they began to seek injunctions against Los Angeles.

With Los Angeles's ability to tap the underground basin potentially limited by the injunctions, the city responded by looking to the overdue construction of a reservoir in Long Valley. Although Fred Eaton still owned the Long Valley site, back in 1905 he had reluctantly consented to a reservoir easement permitting a 100-foot dam. But the original aqueduct plan called for a 140-foot dam, creating a reservoir with a 260,000 acre-foot capacity. The 100-foot dam would impound less than 84,000 acre-feet, or approximately one-third the original plan's capacity.[3] Owens Valley residents believed that such a reservoir was too small to serve the needs of both farmers and city dwellers and if constructed would leave the farmers unprotected. The city nevertheless agreed to release to the farmers their required supply for irrigation and began constructing the dam. But a suit was filed by the farmers to halt construction because of their fear that the reservoir would be insufficient to supply the water the city had promised them. Owens Valley residents also believed that future development of the Bishop region would be severely curtailed because farmers would receive only what they currently needed rather than the amount required if additional land were brought under cultivation. Valley residents insisted that either a 140-foot dam be constructed – which Los Angeles was unable to provide unless Eaton relinquished his Long Valley property – or, as an alternative, that water rights in the valley be adjudicated by the courts before the 100-foot dam was built. Rather than adjudicate the water rights, thus fixing the priority of farmers for all time, the city stopped work on the dam.

The Owens Valley Irrigation District

The city's refusal to permit court adjudication of the water rights resulted, early in 1920, in the resurrection of the idea of an irrigation district for the Owens Valley. The idea had surfaced more than ten years earlier in 1909, and it was recommended at the time as a means of curtailing further encroachment of the city on the Owens Valley's water rights. But lack of interest caused the idea to be dropped. Now, however, as Los Angeles began to consider alternatives for augmenting the aqueduct flow, there was a growing sense of urgency about the need to protect the valley. To gain support for its argument favoring the organization of an irrigation district, the *Inyo Register* published the following call to action:

> We have no organization at all, . . . we are sitting idly by and allowing our rights gradually to be encroached upon.
>
> We must unite every water interest under one organization. When organized we can deal with the city and power companies on an equal footing. . . .
>
> [W]e must organize an irrigation district and we must organize it now, and all of us must join in this district. All the ditches and all the land watered by the creeks flowing into Owens River.
>
> When this is done . . . the waters cannot . . . be taken from our lands either by purchase or in any other manner.[4]

A week later the *Register* again stressed the importance of forming an irrigation district: "It appears to us that the most urgent thing for the people of this valley to do is to further, as rapidly as possible, the organization of an irrigation district . . . [because] the united power of the people backed by the State's laws, will be a formidable force, instead of disunited elements to be antagonized and beaten in detail if the situation should come to that."[5] The campaign continued the following week: "[Knowing that the water] is yours and that you can use it when you see fit," the *Register* advised, "is what gives value to the land and stability to a community. Now, how shall we protect this land? By joining for our protection in an irrigation district, under the laws of California."[6]

The *Inyo Register's* advice fell on deaf ears. Large-scale cooperation was alien to valley residents, as evidenced by their unwillingness to organize a drainage district a year earlier. In August 1921 the *Register* again broached the subject:

> What has become of the irrigation district? Some months ago there was active interest in the formation of an irrigation district to embrace the lands of the northern end of Owens Valley, . . . [but] the whole matter seems to have been completely dropped. . . .
>
> The urgent purpose of an irrigation district when it was proposed was to insure that 'the water should be tied to the land.' . . . Time's passing does not reduce the need for some such step; on the contrary, revelations of what may happen point all the more to a need for some definite system of management of valley water, rather than the everyman-for-himself plan of meeting any emergency or resisting any legal attempt to take away our richest possession. The formation of a district [is] the greatest step that can be taken in self defense.[7]

Still, valley residents took no action. Finally, in an attempt to shake the valley out of its indifference, the *Register* published a front-page map showing the properties in the Owens Valley owned by Los Angeles.[8] The pattern of landholding at the time was not substantially different from the pattern in 1909 (Fig. 8.2), but the map reinforced the call for action, for it graphically revealed how Los Angeles's properties extended unbroken from Owens Lake to a point just three miles southeast of Bishop. Yet even this was not enough to overcome the valley's intrinsic disposition toward inaction.

Meanwhile, Los Angeles desperately needed to supplement its aqueduct supply. The city's only recourse, short of constructing a reservoir in Long Valley that would be large enough to satisfy valley residents, was to launch a new round of purchases of water-bearing lands in the north end of the valley. The first reported incidence of this new policy was in May 1922, when the *Register* glumly noted that these acquisitions were no doubt made "with the idea of stripping them of their water and returning them to the same condition of primitive barrenness that prevails on the other properties [Los Angeles] has secured further south."[9]

This new round of land purchases by the city was the catalyst that finally inspired action on the irrigation district plan. In June 1922 a petition was circulated in the Bishop–Laws–Big Pine region in an attempt to solicit signatures for placing the proposal before the voters. In December the referendum was taken. The proposal needed at least two-thirds of the vote for approval; it received overwhelming support, with 596 voting for the formation of a district and only 27 opposed.[10] Within four months, in April 1923, the Owens Valley Irrigation District was declared one of the public corporations of the state. It included most of the north end of the valley except Round Valley and encompassed

53,390 acres (Fig. 10.1). Its principal function was to act as a bargaining agent that would effect unity of purpose and "mutual protection against encroachments by interests other than agricultural."[11] Water was to be secured by the irrigation district and controlled by the management authority to prevent the piecemeal disassembly of farming by Los Angeles as its thirst for water-bearing lands grew. An ancillary benefit would be to reduce the serious waste resulting from the farmer's current irrigation practices through a unified system of management and distribution.[12]

The Irrigation District's Demise

With Owens Valley residents finally united in their efforts to resist the city's actions, Los Angeles, out of frustration, had to move quickly. In March 1923, while the legal technicalities for perfecting an irrigation district under the laws of the state were being finalized and before water rights could legally be transferred to the new district, the city obtained options on a controlling portion of the lands irrigated by the McNally Canal, amounting to approximately 8,000 acres. Los Angeles, however, quickly realized that no additional water came with these lands, for the water to which the McNally ditch owners were entitled was diverted before it reached the city's aqueduct. The prolonged drought caused the flow of the Owens River to be lower than normal, and farmers both above and below the McNally Canal had rights to more water than they were actually receiving. Below the McNally lands were the lands irrigated by the Big Pine Canal. The elimination of the McNally ditch meant that in the future there would be more water for farms in the Big Pine region.

Even though the farmers of the Big Pine district, whose lands totaled 4,500 acres, signed an agreement to stand together against Los Angeles, the city's affluence overcame this citadel of opposition in October 1923.[13] In an attempt to forestall discouragement among the remaining landowners of the newly formed irrigation district, the following item appeared in valley newspapers:

> In buying this property at Big Pine the City's representatives have again demonstrated their absolute disregard for Owens Valley. . . . When they purchased the majority of the McNally Ditch . . . they found that they could not get this water, and consequently comes the deal for the Big Pine Canal.
>
> [Although these purchases] have probably succeeded in practically mining these two sections as far as future development goes, they have also succeeded in stabilizing the future of the section around Bishop

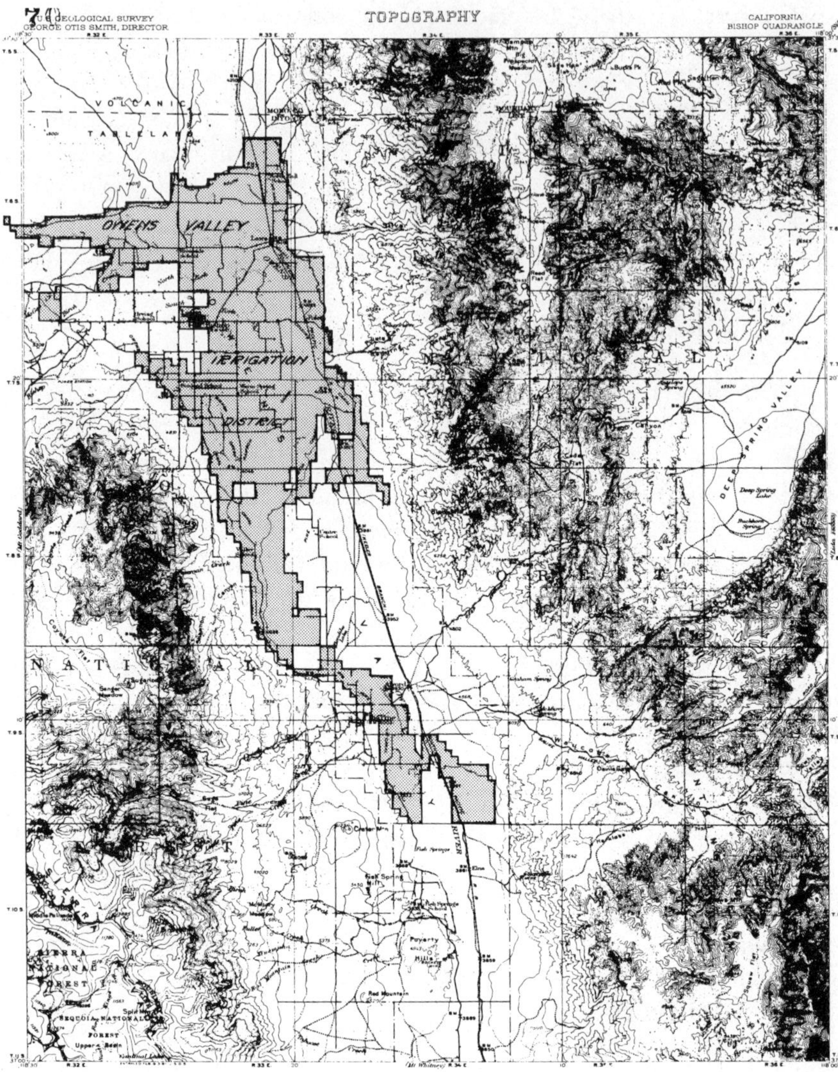

Figure 10.1. The Owens Valley Irrigation District. (Courtesy of Los Angeles Department of Water and Power)

so that in the future this part of Owens Valley will go ahead as it never has before. The water of this part of the valley is safe. The Irrigation District has acquired options on it that makes it sure that the water will soon be turned into the District. Then it will be safe for all time to come.

> Now, if the city would take over the entire McNally Ditch area on the east side of the river, all of the Big Pine Canal area, all of the Rawson Ditch area and all of the Sanger Ditch area, there would still be in this section 33,000 acres of the finest kind of land, and the water rights of this land would be absolutely unquestionable and safe for all time to come. Then when we take into consideration that this land is probably not developed up to 25 percent of its possibilities, one can easily see that this section can still be made more prosperous than it ever has been before–in spite of the purchases of land by the City. . . . [I]n a few years the development of this section will more than offset any loss that has been incurred.[14]

It is true that the majority of stock of the Owens River Canal and the Bishop Creek Canal – the two most important irrigation developments in the Bishop district – had by this time been optioned to the Owens Valley Irrigation District, but Los Angeles had countered by bringing suit to block the sale of bonds by the district. Irrigation districts could secure the rights of existing systems either by purchase or by condemnation. The directors of the Owens Valley Irrigation District believed that condemnation was the least desirable course, so the expense of purchase was to be met by issuing bonds. Because of technical deficiencies in the bond sale, however, in February 1924 the California Bond Commission ruled that the Owens Valley Irrigation District could not use the money from the sale of irrigation bonds for the purchase of canal stock. The district's only recourse, then, was to condemn the remaining ditches.[15] The *Owens Valley Herald* presented a justification for this action:

> With the utter demoralization that seems to be the prevailing passion with many of our people, they have been saved against themselves . . . as . . . these condemnation suits will again protect the land and water owners here until such time as when, if they do sell, they will get for their property what it is worth.
>
> It is not the intention of the Irrigation District to try and prevent the sale of land in this section, if the price offered is sufficient, but it is the intention of the Directors of the District to see that everyone is protected. If a man does not sell he will be absolutely certain that he will receive his water through the operation of the District, but the sale of the property is up to him personally. All the District will do is to guarantee to him that he will not be hurt if he does not sell, and that he will be protected as to price, if he does sell.[16]

But after "seeing the color of the city's money" resulting from the McNally and Big Pine Canal deals, other canal groups decided to sell out as well.[17] By the time the condemnation suits were filed in early March, the city had already dismantled nearly half of the Owens Valley Irrigation District by acquiring 24,759 acres and appurtenant water rights within the original boundaries of the district.[18] The irrigation district had been so mutilated as to make its operation impractical.

The Sacrifice of the Owens Valley

The city was on the move, purchasing farms and appurtenant water rights from anyone willing to sell, causing neighbors to become bitterly divided. Many landowners were uncertain whether to accept or cultivate an attractive offer, while others were determined to hold on to their properties.[19] But the reversion of northern valley lands to desert conditions, with dying trees and dilapidated houses, created a feeling of despair and insecurity among the remaining farm population (Fig. 10.2). Nearby valley towns – Bishop, Big Pine, and Laws – were increasingly deprived of the support of surrounding farms and ranches. Most farm-related businesses were forced to reduce the number of their employees as receipts declined because of the city's extensive purchases.[20] Further, Los Angeles's acquisitions quickly impaired the value of remaining properties, for once the process of land abandonment had begun, there was little outside interest in Owens Valley real estate. By 1924, federal and state land banks were refusing to make loans in the valley. The city's indiscriminate purchases thus demoralized the Owens Valley residents because the transactions left the valley almost helpless and extracted from it nearly all the commercial value that it had secured over time. Even the most stalwart owners were eventually compelled to sell to the city, for they found themselves confronted with a condition of isolation surrounded by desolation, and they realized they could no longer hold on without their neighbors.

The local Indian population was also severely impacted by the city's actions; in fact, they suffered most from the valley's economic collapse.[21] Even though property owners in the valley, whether farmers or businessmen, suffered losses, they at least had the option of selling their property to the city, generally at good prices. But very few Paiutes owned any property; they depended upon wage labor for survival. As in the past, most Paiute men found employment in the fields surrounding Bishop, Laws, and Big Pine, and the women worked in the households in and around those towns. The devastation of local agriculture and the decline in the valley's town population meant that the Indians

Figure 10.2. An abandoned farmstead in the West Bishop region. (Reproduced from Baugh, "Land Use Changes in the Bishop Area," by permission of *Economic Geography*)

were suddenly rendered destitute, with many becoming "objects of charity."[22]

The economy of Owens Valley was now "suspended on a wire of doubt."[23] Residents had lost control of their own lives; they could only respond to the city's actions. Instead of fighting to preserve their way of life, they were more concerned about saving their property interests. A report submitted by the Owens Valley Farm Bureau in July 1924 recommended that either the city purchase the entire valley at a reasonable price fixed by an impartial board of appraisers or that Los Angeles institute a program to reverse the agricultural destruction of the valley. "Why should it be necessary," questioned the bureau, "for the city to buy out a few people scattered here and there and let the others suffer severe depreciation through the reverting of these lands to the wild state?"[24]

By now the majority of Owens Valley residents favored the wholesale purchase plan. Accordingly, the remaining privately owned lands in the north end of the valley were offered for sale as a package for $8 million, a figure that also included reparations for the loss of business in the towns. Some farmers, because of their love of the land, opposed sale at any price,[25] but most valley residents were now reconciled to the fact that they were competing with the city over the valley's water supply in a contest the valley could not possibly win.[26] Considering the consequences resulting from the city's initial purchases, the only fair solution, it seemed, was to buy the whole district. Los Angeles, however, initially opposed the wholesale purchase idea. In fact, in October 1924 the city halted its purchase program and drew up a plan to guarantee sufficient water supplies for the irrigation of approximately 30,000 acres in the north end of the Owens Valley. Nearly half of this total acreage was located in Round Valley, however, meaning that the amount of irrigated land around Bishop would be significantly reduced. Distrustful of the city's motives and convinced that without storage a supply necessary to water 30,000 acres could not be assured during dry years, the valley refused the offer and stood by its original proposal of a wholesale cash settlement. In January 1925 the city withdrew its irrigation proposal and resumed its random land purchase program in pursuit of the piecemeal destruction of the Owens Valley. According to W. A. Chalfant, the Owens Valley was relegated to a status unique in California—that of facing a hopeless future as it reverted to "primitive waste."[27]

By 1928, Los Angeles had secured approximately 90 percent of all water-bearing lands in the Owens Valley. The additional property in the north end of the valley not only enhanced surface flow into the aqueduct but it also significantly expanded the area for pumping. However, the few remaining farmers in the valley possessed claims to water superior to those of Los Angeles, and they could conceivably bring legal action against the city's pumping any time the reasonable use of their subterranean water rights was affected.[28] To avoid potential water conflicts, in the spring of 1929 Los Angeles adopted a policy of purchasing all the remaining land under private ownership in the Owens Valley, including town properties if owners wished to sell them. This policy allowed the city to avoid paying reparations to citizens of towns who lost business as a result of the farm purchases.[29] An Act of Congress followed that protected the watershed supplying the Los Angeles Aqueduct by withdrawing the remaining public lands in the valley to prevent speculative entries ensuing from the city's "purchase all" program. By 1933, Los Angeles had secured 95 percent of all water-bearing parcels in the valley and 85 percent of the town property.

Los Angeles still faced another problem. The increasingly desperate plight of the Owens Valley Paiute was a potential source of additional negative publicity. In 1902 a portion of the former military reserve at Camp Independence had been turned over to the Paiute for the establishment of the Fort Independence Indian Reservation. Those Paiutes who elected not to live on the reservation and who owned no land led "a precarious existence, depending almost entirely on the goodness of their white neighbors or friends of their race for shelter."[30] Soon after the withdrawals that the Forest Service had made on Los Angeles's behalf were lifted in 1911 and the Owens Valley was reopened to settlement, the Department of the Interior granted nearly 69,000 acres of widely scattered sites for the use of the homeless Paiutes. The largest tract encompassed 67,164 acres of barren ground on the volcanic tableland north of Bishop, and the remainder was scattered throughout the Owens Valley.[31] With the collapse of the valley's economy, conditions among the Owens Valley Paiute worsened. By 1935 only 35 percent of the Paiute households had one or more persons employed full- or part-time, compared with 83 percent in 1880.[32] As farmers, ranchers, and townspeople emigrated, many more Indians were left homeless, a situation that led the city to advocate their removal from the valley.[33] The city's preference for relocation, however, met with resistance, this time from both the federal government and the Indians themselves.[34] In order to maintain a low profile, Los Angeles began to look for alternatives to Indian removal.

The city developed a suitable plan that solved its dilemma and simultaneously improved upon the Indian land base, in quality if not in quantity. At the time, Paiute lands were dispersed among Los Angeles's holdings, thereby hindering the city's attempts to consolidate its control of the Owens River watershed.[35] In addition, the Los Angeles Department of Water and Power was legally bound to supply water to the Indian homesites that held water rights under federal law. This reduced the potential water supply that might flow into the aqueduct. The solution to the problem came when the Owens Valley Land Exchange was approved by an Act of Congress in April 1937 and by the Department of Interior in June 1939. The agreement involved a trade of 2,914 acres of scattered Paiute homesite lands with appurtenant water rights for 1,392 acres of prime city-owned land in the valley with a guaranteed water supply for the establishment of three new homesite areas.[36] The largest, the Paiute-Shoshone Indian Reservation, was located west of Bishop on 875 acres, while 280 acres and 237 acres were used to create the Big Pine and Lone Pine Indian Reservations, respectively. The Indians living on 360 acres at the Fort Independence Reservation were offered an additional 120 acres. However, because they already possessed high-quality land and were suspicious of the city's intentions, they elected to opt out of the agreement. The federal government agreed to supply the new homesites with sewer systems, efficient irrigation, and adequate housing.[37] Currently there are approximately 1,500 Paiutes and Shoshones living on the four reservations. On the one hand, the land exchange provided the Paiute with superior lands with a guaranteed water supply and proximity to the amenities and labor opportunities in the valley's towns. On the other hand, the city managed to achieve greater efficiency in fulfilling its obligation to supply the Indian lands with water. Consequently, the Owens Valley Land Exchange was remarkably successful in balancing the needs of the city and those of the Owens Valley Paiute.[38]

Today, with the exception of these Indian lands and a few other scattered holdings, virtually all the land in the Owens Valley is owned by Los Angeles. Los Angeles also went on to acquire the Long Valley reservoir site and proceeded to construct the Long Valley dam, creating Crowley Lake in 1941. Today Crowley Lake stands as a memorial to the city's utter disregard for the needs of valley farmers, for it was completed after nearly all private holdings in the Owens Valley came under Los Angeles's control. A more timely construction of this reservoir could have been a pivotal step toward obviating the need to sacrifice the rural livelihood of the valley for the domestic requirements of Los Angeles.

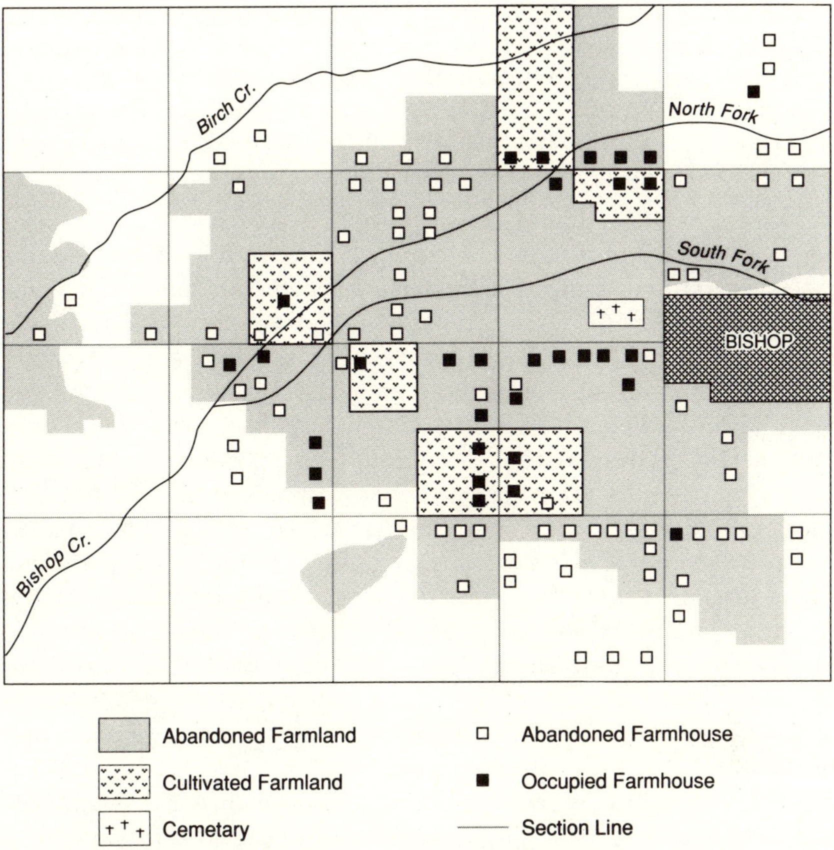

Figure 10.3. Abandoned farmsteads in the West Bishop district in 1935. These lands were included in the Owens Valley Land Exchange of 1937 and ultimately were brought back into cultivation by the Owens Valley Paiute (see Fig. 5.4). (Adapted from Baugh, "Land Use Changes in the Bishop Area"; redrawn from Sauder, "Owens Valley's Abandoned Landscapes," 65)

Abandoned Landscapes

A unique form of urban blight settled over the Owens Valley as most of its productive farmland, no longer irrigated, was abandoned. The number of farms in Inyo County dropped from 521 in 1920 to 201 in 1935 as the amount of irrigated acreage declined. By 1935 the destruction of agriculture in the Bishop region was nearly complete (Fig. 10.3).[39] Geographer Ruth Baugh, writing in the mid-1930s, described the changes that had occurred there since 1916:

> From the railroad station at Laws to the town of Bishop and west to Red Hill one traveled [in 1916] tree-lined roads through a district almost continuously cultivated; green fields bordered by wide irrigating canals alternated with stretches of damp pasture lands. A verdant land, Owens Valley presented a scene of substantial economic well-being and human contentment. How different today the area, vacant and desolate![40]

Baugh also noted that there was little evidence in the town of Bishop, the valley's largest service center, that grain, hay, fruit, honey, livestock, dairy, and poultry products had once been exported from the valley.[41] Laws had lost its critical mass for survival by this time and had largely ceased to exist. Bishop and Big Pine, as well as Independence and Lone Pine, however, were saved from extinction by the growing number of tourists taking advantage of the eastern Sierra's scenic resources.

Dying orchards, empty schoolhouses, and abandoned farmhouses and farm buildings became common landscape features. Los Angeles, the new landlord of the Owens Valley, removed most traces of abandoned rural life as quickly as possible; houses and barns were bulldozed or burned and trees and orchards were either cut down or uprooted in order to hide from passing motorists the despoliation that it had wrought. Nevertheless, subtle landscape features survive to remind us of the hardship and toil its former residents endured to make this arid land bloom.

Viewed from the air, relict field patterns are etched into the floor of the Owens Valley (Fig. 10.4). Their rectangular shape was influenced by the public land survey of the mid-1850s and the nineteenth century public land measures designed for disposal of the public domain, but their expanse reflects the breadth of farming and ranching that once prevailed in the valley (Figs. 10.5 and 10.6). The resolution of conflicts between stockmen and farmers demanded fences, but in a treeless region like the Owens Valley, wooden rails were expensive and difficult to obtain. Some early farmers therefore encircled their fields with substantial stone walls (Fig. 10.7). Although there was little time for extensive tree culture to furnish fencing materials, the black locust, lombardy popular, and cottonwood trees pioneer settlers planted along roads and as windbreaks around fields and farmhouses have survived to the present and still ward off the valley's heat and aridity (Fig. 10.8).

Many of the late-nineteenth-century improvements in the northern half of the Owens Valley were linked to congressional approval of the Desert Land Act in 1877, and the dry irrigation ditches that wind across the valley floor attest to one of the few cooperative enterprises

Figure 10.4. Relict field patterns south of Bishop. (Courtesy of Los Angeles Department of Water and Power)

valley farmers attempted (Fig. 10.9). But Los Angeles's acquisitions caused land formerly opened to farming by the expanding network of canals to be abandoned, and withering trees and the invasion of rabbit brush often mark the location of former farmsteads (Fig. 10.10). Many of the abandoned farms in the Bishop region were engaged in dairy production, and the remains of Roberts's silo, the first to be constructed in the Owens Valley, are still on view (Fig. 10.11). The former importance

Figure 10.5. Former farmland surrounding Bishop in 1931. (Courtesy of the UCLA Department of Geography Aerial Photo Archives)

Figure 10.6. Former farmland near Laws in 1931. (Courtesy of the UCLA Department of Geography Aerial Photo Archives)

Figure 10.7. A relict stone fence in Round Valley. One Round Valley farmer, John Birchim, used stone walls extensively to encircle his fields.

Figure 10.8. Lombardy poplars planted by pioneer settlers still provide tree-lined drives in the West Bishop region.

Figure 10.9. The dry channel of the Owens River Canal west of Bishop.

of dairying in the Bishop–Round Valley region is reflected in today's landscape in the numerous abandoned concrete silos that stand forlorn against the mountain backdrop (Fig. 10.12).

There are also more obvious remnants of past rural activities in the Owens Valley. Los Angeles leased back a portion of its holdings to farmers, so agriculture has not been completely expunged from the valley floor. Unlike former times, however, Los Angeles gained priority in the use of the Owens River, and prior to 1970 ranchers received city water on a precarious basis – whatever water the city could afford to relinquish in any given year. Because of this uncertainty, farming today is at a mere fraction of its former level, and instead of orchards and expansive green fields, short-term leases have limited rural activities to stock raising and small-scale feed crop production (Figs. 10.13 and 10.14).

Alfalfa, beef, fruit, and dairy production were the chief agricultural enterprises in the Owens Valley when Los Angeles embarked on its final acquisition program and abruptly terminated the evolving rural livelihood patterns in the valley. Each activity was a small-scale endeavor

Figure 10.10. Dying trees and an invasion of rabbit brush mark the location of a former farmstead in the Warm Springs district southeast of Bishop.

compared to other agricultural districts in the state, primarily because of the valley's relatively short growing season and its persistent isolation from the more settled regions of California. Yet much of the historical ambiance of the Owens Valley evolved out of its farming and ranching activities, and the valley's abandoned rural landscapes serve as reminders of the hardships associated with settling this arid region.

The Owens Valley's abandoned landscapes also dramatically demonstrate how the shift of water away from farming can have significant implications for those who are tied to agriculture and related activities. The Owens Valley is a lost frontier today, no longer the home of self-reliant farmers and ranchers occupying one of the earliest regions of rural development in the western Great Basin. Instead, it has become a tributary province of the city that it helped to build.[42] Since the West's water supply is largely appropriated today, and water reallocations will

Figure 10.11. The remains of Roberts's silo in Round Valley. Constructed of light tufa rock, it stood twenty-two feet high and held 230 tons of silage.

Figure 10.12. Abandoned concrete silos located southeast of Bishop. Many silos, although unused today, are still seen in the Bishop-Round Valley region, where dairy farming was once prominent.

Figure 10.13. An irrigated pasture in Round Valley. In contrast, most of the valley floor is used today as rangeland for livestock.

therefore inevitably increase in importance, the abandoned landscapes of the Owens Valley should stand as reminders of what could happen to small communities elsewhere in the region as rural-to-urban water transfers become more common. William Kahrl, in his book *Water and Power*, asserts that the "ethic of growth" laid waste the Owens Valley.[43] Without proper safeguards, the same ethic will doubtless sweep across the arid West in future years to engulf other rural communities.

Figure 10.14. A working alfalfa ranch near Red Mountain south of Big Pine. Alfalfa is now grown on only about 2,800 acres of valley land.

11. Water Transfers and Rural Preservation

> One must ask whether anything that might accelerate the increasingly malignant growth of Tucson, Los Angeles, Denver, and San Diego can be considered a good thing, regardless of other possible benefits that might follow.
>
> —James R. Udall, 1987

Rural-to-Urban Water Transfers

The Owens Valley episode is now deeply rooted in the cultural memory of rural water users throughout the West because it illustrates what can happen when water transfers are allowed to operate without adequate safeguards. Although Owens River water began flowing into the Los Angeles Aqueduct eight decades ago, conflicts between the city and Owens Valley residents have persisted to recent times. With the West's water supply now largely appropriated, in future years agricultural communities throughout the region may be forced to protect their way of life in the face of mounting demands for water by the region's growing urban areas. This concluding chapter employs the Owens Valley experience as a vehicle for exploring avenues for facilitating voluntary water transfers without imposing undue stress on rural communities.

Currently, agriculture accounts for 85 to 90 percent of the water used in the arid West, and incessant migration to Western cities, coupled with declining federal interest in financing huge water projects, ensures that pressures for water transfers out of agriculture will increase as the competition for existing supplies intensifies.[1] Western municipalities, in fact, now purchase distant water rights as a general policy in order to mitigate the risk of future water shortages while also providing a base for economic growth.[2] The approval of new housing developments around Phoenix and Tucson, for example, is contingent upon proof that long-term water supplies are available to support this growth.[3] The rapidly growing Front Range communities in Colorado also purchase water long before they intend to use it. Water rights to more than 60,000 acres of irrigated land in the Arkansas River Valley have already been sold to these communities.[4] Utah's growing demand for energy, coupled with its stock of energy resources, is creating substantial demands for the transfer of water out of agriculture.[5] And Las Vegas, the fastest growing major city in the country during the decade of the 1980s, recently filed for unappropriated groundwater in nearby rural counties, causing farmers and ranchers to fear that their land will be dried out by the city's pumping.[6] Newspaper headlines proclaiming "Battle Lines Drawn in Sand as Las Vegas Covets Water" and disturbing reports of how urban legislative interests in California have "banded together to oppose agriculture" suggest that history may be about to repeat itself.[7]

But the Owens Valley experience also illustrates that more than a region's economic base and community cohesion can be affected by rural-to-urban water transfers. It is now clear that the removal of substantial quantities of water from the Owens Valley hydrological basin has significantly impacted the valley's ecology. Although colonization was responsible for the initial environmental changes – including the reduction of natural bunch grasses due to overgrazing by livestock, the clearing of sagebrush lands for agriculture, the conversion of Indians' irrigated meadows to alfalfa and grain fields, the saturation of farmland due to over-irrigation, and the reduction of Owens Lake – the environmental alteration of the Owens Valley was accelerated following Los Angeles's intrusion.[8] Completion of the Los Angeles Aqueduct in 1913 left a 50-mile stretch of the Owens River dry below the intake (Fig. 11.1), and by 1924 Owens Lake was reduced to a dry alkaline sink. Prior to the diversion of the lower Owens River into the aqueduct, the river channel supported a riparian habitat of cottonwoods and tree willows, and lush salt-grass meadows were distributed along its bottomlands. Carved into the floodplain was an extensive system of relic meander channels that supported a dense growth of reeds, rushes, and cattails. This rich natural ecosystem furnished cover for numerous bird species. The capture by the aqueduct of the lower portions of streams tributary to the Owens River resulted in the loss of additional riparian habitats and meadow areas as well. The aqueduct ultimately was responsible for the diversion of 10 percent (in stream miles) of the Owens River and its tributary streams out of their channels, and the vegetation along these sections either died off or has been greatly reduced.[9] Finally, dust generation at the surface of the dry bed of Owens Lake has resulted in severe air-quality problems in the Owens Valley.[10]

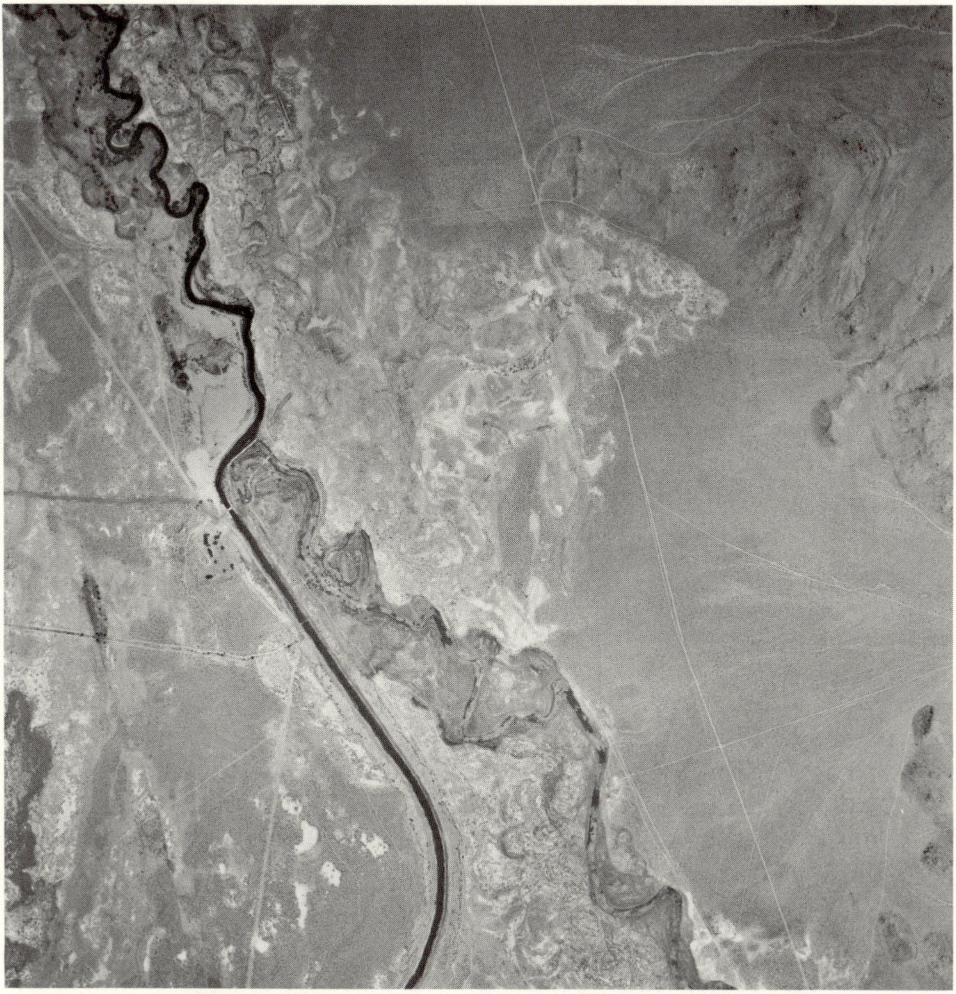

Figure 11.1. An aerial photo of the unnaturally straight channel of the Los Angeles Aqueduct Intake at left and the meandering channel of the lower Owens River to the right. The portion of the river below the intake recently began receiving water to repair the riparian habitat. (Courtesy of Los Angeles Department of Water and Power)

Even though the Owens River ran dry south of the intake following the completion of the Los Angeles Aqueduct, the environmental issue did not emerge in the valley until after 1970, when Los Angeles opened a parallel aqueduct to supplement the first. Completion of the second aqueduct, which was supplied by increased groundwater pumping, allowed the city to augment its water export from the Owens Valley by 60 percent.[11] Los Angeles had been periodically pumping groundwater from the Owens Valley water table for many years with little noticeable effect, but the city now began to implement a systematic program of extracting groundwater. The potential impact of the city's stepped-up pumping quickly came under scrutiny in the valley. The concern was that heavy pumping would so lower the water table that the natural vegetation would die as the valley's springs, meadows, and marshes dried up. Indeed, the increased pumping caused large fluctuations in groundwater levels and a significant drawdown of the water table in certain areas. It is estimated that Los Angeles's water-gathering activities since 1970 have adversely affected 25,000 acres of groundwater-dependent vegetation in the valley.[12]

In 1972 Inyo County filed suit enjoining Los Angeles from further pumping until water-table levels that would not be harmful to the valley's vegetation communities could be determined. Los Angeles pointed out, however, that it had secured the rights to both surface and groundwater legally and permanently, and that a 1963 Supreme Court ruling granting Arizona and Nevada increased rights to Colorado River water necessitated additional exports from the Owens Valley to compensate for this loss.[13] Inyo County voters responded to Los Angeles's increased pumping by passing a groundwater law that imposed new restrictions on the city's pumping. Los Angeles challenged the law in court, and both the city and Inyo County pressed legal threats against each other until a recent out-of-court agreement was reached.[14] In essence, the pact ensures that Los Angeles can continue its groundwater pumping while simultaneously promising environmental guarantees to Owens Valley residents. In other words, no longer can the water needs of Los Angeles take priority over the Owens Valley environment. More enhancement projects are also to be implemented to compensate for past environmental damage. For example, Los Angeles is currently attempting to mitigate some of its activities by restoring the flow of the lower Owens River and by revegetating severely impacted areas in the valley with native species. Alfalfa has been planted in other areas in an effort to lessen the blowing dust resulting from the reduction of vegetation cover. But Los Angeles's wholesale land and water acquisitions in the Owens Valley created an anomalous situation whereby the city became the owner of virtually the entire valley, and the city has often faltered in its "attempts to come to terms with its responsibilities under this essentially colonial relationship."[15] In other words, Los Angeles has tended to give less than due consideration to moral and ethical issues concerning the valley's environment, and it generally has not yielded to environmental constraints until forced to do so by law.[16] Los Angeles resorted to compromise only recently when its legal powers proved to be much less than it had hoped.[17]

Prior Appropriation

To appreciate fully the complexities associated with the water transfer issue, it is useful to review early water resource developments in the Owens Valley and in the West generally. Although mining contributed significantly to the initial economic growth of the western United States, including the Owens Valley, it was the snow-fed streams flowing off the higher mountain slopes that made permanent settlement possible and that shaped the early patterns of occupancy in the region. The fragmented pattern of irrigation and settlement that emerged in the intermountain West attests to the symbiotic relationship that developed between mountains and lowlands in this arid region (Fig. 11.2).[18] Irrigated farming, therefore, was primarily responsible for the West's sustained economic growth, and the demand for irrigation led to the region's water laws and institutions. Borrowing from the custom of Western miners' use of water, pioneers adopted the doctrine of appropriation, or the right to appropriate water on the public domain, which has subsequently proved conducive to long-distance rural-to-urban water transfers in the region.[19]

It was the smaller streams of the arid West, such as those flowing off the Sierra into the Owens Valley, that were first appropriated because they could be diverted to pioneer farms by individual effort with little difficulty. But only a small fraction of the Owens Valley's, and the West's, irrigable lands could be redeemed in this fashion; the diversion of larger streams, such as the Owens River, required either cooperative labor or substantial capital investment. Approval of the Desert Land Act in 1877 inspired the construction of a number of irrigation canals in the Owens Valley, most of which were fed by the Owens River and constructed on a cooperative basis. The partnerships built comparatively small ditches that irrigated only the valley floor, leaving the higher alluvial slopes unwatered. This circumstance prevailed throughout the dry West and led to congressional approval in 1902 of the Newlands Reclamation Act, whereby the federal government would assist in the development of large irrigation projects in the region.

Approximately one-quarter of the surface water used for irrigation in the West is diverted independently by farmers, such as those who

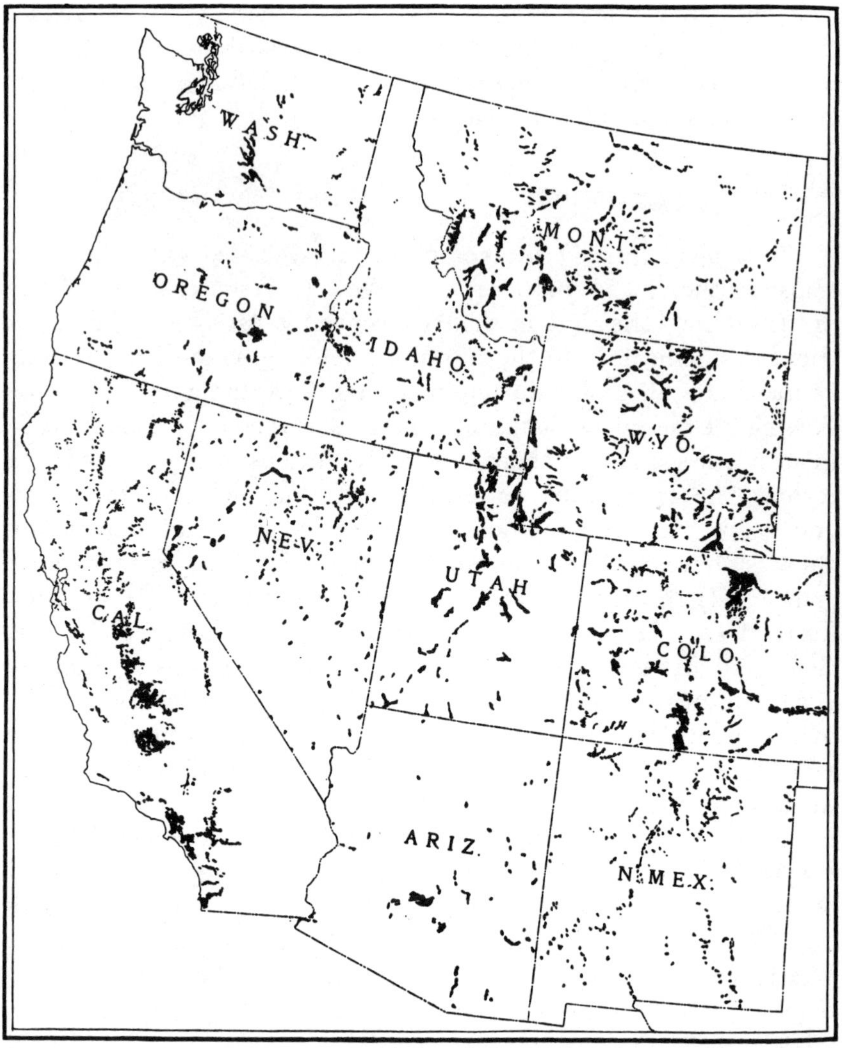

Figure 11.2. The fragmented pattern of irrigated lands in the western United States in 1899. (Reproduced from *Twelfth Census of the United States Taken in the Year 1900: Agriculture – Part II, Crops and Irrigation,* 802)

arrived first in the Owens Valley, while the remaining 75 percent is diverted by some type of irrigation organization.[20] Nearly all of the land irrigated by organizations is owned by private mutual irrigation companies similar to those that developed in the Owens Valley following approval of the Desert Land Act or by larger public irrigation districts.

About 15 percent of the West's water is supplied by federal projects.[21] In the context of potential water transfers, irrigated agriculture is important not only because of the quantity of Western water it consumes but also because irrigators hold a high proportion of senior appropriative rights in the region. If future water transfers are to bring water use in the region into greater balance, irrigators must relinquish some of their senior appropriative rights to the new higher-value uses. The method by which water rights were lost in the Owens Valley exhibits the need for institutional safeguards so that water transfers may proceed without devastating the economic, social, and environmental bases of rural communities.

Lost Opportunities

Los Angeles's inability or unwillingness to make its intentions clear in the Owens Valley is largely responsible for the undesirable consequences. It began in 1905 when the city secretly secured options on land and water rights in the southern half of the valley, undermining a potential reclamation project. Further, it failed to build a major reservoir above the aqueduct intake that would have allowed for sufficient storage and prevented the water crisis of the 1920s. Finally, when the water famine did occur, it pursued a policy of buying whatever properties it could without accepting the responsibility for the effect of its actions on either the economic life or the morale of the Owens Valley community. Yet inaction among northern valley residents also contributed to the valley's demise, for its citizens adopted no coordinated plan of their own to counter a Los Angeles intrusion that was certain to materialize, given the city's phenomenal growth and inadequate local water resources. Although valley residents belatedly attempted to organize an irrigation district to deter Los Angeles from sacrificing the Owens Valley's future for that of its own, their actions came too late to reverse the course that the city was determined to follow. The Los Angeles–Owens Valley case demonstrates that initial water rights purchases in a region increase the probability of future transactions. City governments, concerned about costly litigation and damage claims, will no doubt attempt to buy out all existing water users, as happened in the Owens Valley.

In retrospect, the devastation of agriculture in the Owens Valley and the environmental impact associated with the water transfer could have been prevented. Carefully planned and located reservoirs, for example, are generally desirable alternatives to the piecemeal disassembly of agriculture as water is reallocated for urban uses. Although most rural communities that depend on irrigation have provided for storage, the Owens Valley had no storage facilities when Los Angeles began to covet Owens River water. In its rush to get water to the city, Los Angeles simply diverted the unappropriated surplus water flowing in the river channel. The decision not to construct a large storage reservoir in Long Valley was the primary reason for the eventual sacrifice of agriculture in the Owens Valley. At the time of the aqueduct diversion, average runoff from the Owens River watershed during normal years was more than enough to supply both the valley and the city. But without a reservoir, there was no means of storing the precipitation of wet years, and when the dry years came, there was insufficient water to supply both the city and the valley. With proper storage facilities, systematic water exchanges between municipalities and agricultural ditch companies can be effected.

The Owens Valley scenario also illustrates that a general apathy about water and its use prevails in rural areas unless a crisis situation develops. This laissez-faire attitude toward water places a significant constraint on the ability of rural communities to plan for future water needs. Only when the northern end of the Owens Valley was threatened by Los Angeles's purchases in the early 1920s did valley farmers agree to the formation of an irrigation district that would combine the various ditch companies under a single water management authority. Had the irrigation district been created earlier, the northern valley farmers and ranchers could have dealt with Los Angeles from a position of strength. Water could have been tied to the area encompassed within the proposed irrigation district and controlled by the management authority to prevent the piecemeal disassembly of farming by Los Angeles as its thirst for water-bearing lands grew. But the efforts of Owens Valley farmers to protect their water supply came too late. If residents in other areas tie water to irrigation service areas and adopt creative and flexible measures to allow for some transfers, they can preserve agriculture and its associated way of life. With proper water management and storage, opportunities exist to work within the current institutional framework to implement exchanges.

However, since many of the water institutions that adequately served the West during its developmental phase are no longer suitable under present conditions, certain legal reforms are also necessary to facilitate the more efficient use of water, thereby augmenting the volume of water involved in potential transfers. When supplies could be provided to new users without harming existing ones, the traditional approach to water allocation worked well. Prior appropriation rights were acquired simply by diverting water from a stream, carrying it to the desired location, and putting it to some state-approved beneficial use. Outmoded beneficial-use provisions now tend to inhibit water transfers and incentives to conserve; irrigators are often unwilling to transfer water rights temporarily to other parties, fearing that the transfer will be construed as evidence that they can no longer put their water to beneficial use. In such cases, rights to the transferred water could be lost. Moreover, under current laws the adoption of more efficient irrigation methods does not necessarily guarantee that users will have more water to sell or even to use themselves. The water saved could legally be declared in excess of the amount needed for beneficial use and forfeited. Consequently, it is common practice in the West for farmers to over-irrigate or otherwise waste water that they have been allotted rather than risk losing it through non-use, and this lack of a conservation incentive in the "use it or lose it" provision of beneficial use has led to inefficient water distribution.[22] If all the options for sharing agricultural water supplies are to be explored, institutional reforms will be required to encourage water conservation among rural users.

In fact, the irrigation practices found today throughout much of the West are not significantly different from those used by Owens Valley farmers at the turn of the century. Even though the extravagant use of water among valley farmers was highly publicized, residents never gave serious consideration to the implementation of water conservation projects in the valley, with the water saved being used to supplement Los Angeles's supply. In 1976 the U.S. Soil Conservation Service estimated that total annual water waste resulting from irrigation amounted to 24 million acre-feet per year, a figure that exceeds the total volume of water consumed by all municipalities and industries in the nation.[23] Since large percentage increases in water consumption for nonfarm uses can be accommodated with small relative reductions in agricultural uses, and since wasteful irrigation methods are still employed in many rural areas of the West, a major source of new water is embodied in the conservation of water in existing uses.

In most Western states it is not clearly lawful for a user to sell or lease water that has been salvaged from waste or conserved by making one's operation more efficient. In California, however, the legislature has acted to clarify the principle that one may sell conserved water.[24] In fact, the Metropolitan Water District of Southern California recently negotiated an arrangement with the Imperial Irrigation District concerning water conservation measures. In 1984 the State Water Resources

Control Board ruled that the Imperial Irrigation District's use of water was unreasonable and that conservation measures should be implemented immediately.[25] The Metropolitan Water District was interested in the conserved water because it had been taking Colorado River water allocated to, but not used by, Arizona until the Central Arizona Project began operation in November 1985. Metropolitan's agreement with farmers in the Imperial Valley involves financing conservation efforts in exchange for the water farmers save.[26] A decrease in agricultural diversions to the Imperial Irrigation District from the Colorado River via the All-American Canal will allow a corresponding increase in Metropolitan Water District diversions upstream at the Colorado River Aqueduct. This arrangement illustrates the possibility of urban assistance with rural conservation efforts and the subsequent reallocation of the salvaged water for urban use. At the same time, however, a concerted effort to conserve water among municipalities is also necessary. Had conservation alternatives been applied in the Owens Valley, along with the development of appropriate storage facilities and exchange agreements, the rural devastation of the valley could have been prevented.

No More Owens Valleys

The Los Angeles–Owens Valley conflict is instructive, for it provides a warning about the ability of outlying rural communities in the arid West to protect their economic and social identities, as well as their environment, from the onslaught of urbanization. Since the agricultural and ranching industries and accompanying rural and small-town lifestyles provided the genesis for economic development in the Owens Valley and much of the West, the value structure and economic well-being of such rural communities rest on the maintenance of agriculture.[27] It is evident from the above discussion that irrigated agriculture can readily be "cannibalized" by reliance on unregulated water transfers. If rural-to-urban water transfers are to play an important role in future Western water decisions, it is imperative that Western states develop enlightened water reallocation policies. There are many options for sharing the larger agricultural water supplies with urban users, and it is in the long-term interest of agriculture that more flexible water institutions be developed to facilitate water transfers.

The Owens Valley is unquestionably the country's most celebrated example of how complete freedom for cities can threaten the very existence of some conveniently situated rural communities that happen to be well endowed with water. Local economies and lifestyles in the

American West have been built on a reliance on existing water use and delivery patterns, and the Owens Valley experience illustrates the potential for unrestrained water transfers to devastate agricultural communities. It is apparent that "economic Darwinism" will predominate in the West if water transfers are allowed to take place without proper oversight.[28] Fortunately, today nearly all the Western states have some agency with discretionary authority to oversee proposed transfers, and those contemplating water exchanges now face what James Udall has called "a daunting array of legal, logistical, and financial hurdles."[29]

Nevertheless, water reallocations in the West will inevitably increase in importance, and the Owens Valley's fate illuminates numerous lost opportunities to which other rural communities might look as rural-to-urban water transfers become more common. Even though one must question the wisdom of a policy that enables large metropolises in the arid West to continue to grow larger, the impetus for urban expansion in the region apparently will not be denied. Since much of the ambiance of the West derives from its farming and ranching base, a sound and functional transfer process is critical to the region's continued vitality; otherwise, large cities will ultimately squeeze the water and thus the life out of small communities, as happened in the Owens Valley.[30] By reforming outmoded water laws to encourage increased efficiency and conservation efforts in both rural and urban settings, and by devising creative storage and exchange agreements that are regulated and coordinated at the state and federal levels, water reallocations from agricultural to urban uses can be implemented without effacing the region's tangible history by irreparably harming rural communities. With adequate safeguards in place, there need be no more Owens Valleys, and the ambiance and lifestyle of Western rural areas can be preserved in tandem with the ongoing urbanization of the region.

Notes

Chapter 1. Colonizing an Arid Frontier

1. D. W. Meinig has produced several classic geographical works on the colonization of different regions. See, for example, *On the Margins of the Good Earth: The South Australian Wheat Frontier, 1869–1884*; "The Mormon Culture Region: Strategies and Patterns in the Geography of the American West"; *The Great Columbia Plain: A Historical Geography, 1805–1910*; and *The Shaping of America: A Geographical Perspective on 500 Years of History*, vol. 1: *Atlantic America, 1492–1800*, and vol. 2: *Continental America, 1800–1867*.

2. The region of the western United States located between the Wasatch Mountains of Utah and the Sierra Nevada of California is often referred to as the Great Basin. It comprises the northern half of the Basin and Range Province and consists of linear, roughly north–south mountain ranges separated by valleys, many of which are closed basins. See Hunt, *Natural Regions of the United States and Canada*, 390.

3. Worster, *Rivers of Empire*; Reisner, *Cadillac Desert*.

4. See, for example, Arrington, *Great Basin Kingdom*; Fox, "The Mormon Land System"; Hunter *Brigham Young the Colonizer*; Meinig, "The Mormon Culture Region"; Nelson, *The Mormon Village*; and Thomas, *The Development of Institutions Under Irrigation*.

5. The series of articles by Bowen is composed of "Elko County's Dry Farming Experimental Station," "Turnover of Pioneers and Property in a Marginal Nevada Farming Community"; "Jackrabbit Invasion of a Nevada Agricultural Community"; "Dryland Homesteading on Tobar Flat"; "Pioneer Success and Failure in a Northeastern Nevada Valley"; "The Heritage of an Empty Land: Independence Valley, Nevada"; "The Desert Homestead as a Non-Farm Residence"; and "A Backward Step: From Irrigation to Dry Farming in the Nevada Desert."

6. Patterson et al., *Nevada's Northeast Frontier,* 493–94.

7. Kersten, "The Early Settlement of Aurora, Nevada, and Nearby Mining Camps"; Patterson et al., *Nevada's Northeast Frontier.*

8. Hornbeck, "The Far West, 1840–1920," 268.

9. The debate surrounding how the City of Los Angeles acquired water rights to the Owens Valley is a story filled with political intrigue involving numerous participants. It has been told on many occasions and from a variety of viewpoints. For an overview of the Los Angeles–Owens Valley controversy, see Hoffman, *Vision or Villainy;* Kahrl, *Water and Power;* Nadeau, *The Water Seekers;* and Walton, *Western Times and Water Wars.* The following list of references, although not exhaustive, provides additional insight into the controversy: "A Theft in Water," pts. 1–8 in *Inyo Magazine* (September 1–December 15, 1908), and pts. 9–10 in *Sierra Magazine* (January–February 1909); Anderson, "Are Government Officials Playing Fair? Some Facts About the Owens Valley Condition"; Ibid., "Oppressive Use of Power"; Ibid., "Why Was the Owens River Project Abandoned?"; Chalfant, *The Story of Inyo,* 337–411; Faulkner, "Owens Valley: Where the Trail of the Wrecker Runs"; Hoffman, "Joseph Barlow Lippincott and the Owens Valley Controversy: Time for Revision"; Ibid., "Origins of a Controversy: The U.S. Reclamation Service and the Owens Valley–Los Angeles Water Dispute"; Ibid., "Fact or Fiction in the Owens Valley Water Controversy," 179–91; Kahrl, "The Politics of California Water: Owens Valley and the Los Angeles Aqueduct, 1900–1927"; Kinsey, *The Water Trail,* 7–39; Mayo, *Los Angeles,* 221–46; Moody, "Los Angeles and the Owens River"; Nadeau, "The Water War"; Newell, "The Reclamation Service and the Owens Valley"; Reisner, *Cadillac Desert,* 54–107; Stegner, "Water and Power"; "The Owens Valley Controversy, By a Special Correspondent"; Yonay, "How Green Was My Valley"; Walton, "Picnic at Alabama Gates"; and Wood, *The Owens Valley and the Los Angeles Water Controversy.*

10. One study in the 1930s examined the impact of the Los Angeles Aqueduct on a portion of the Owens Valley. See Baugh, "Land Use Changes in the Bishop Area of Owens Valley, California." Kahrl, *Water and Power,* 32–39, and Walton, "Picnic at Alabama Gates," 194–98, provide brief descriptions of conditions in the valley prior to the water diversion.

Chapter 2. In the Shadow of the Sierra

1. Hill, *Geology of the Sierra Nevada,* 69.

2. Pakiser et al., *Structural Geology and Volcanism of Owens Valley,* 35; Hill, *Geology of the Sierra Nevada,* 169–70.

3. Hill, *Geology of the Sierra Nevada,* 169.

4. Pakiser et al., *Structural Geology and Volcanism,* 1.

5. Ibid., 10.

6. Ibid.

7. Ibid.

8. Lee, *Water Resources of a Part of Owens Valley,* 11.

9. Ibid., 7.

10. Austin, *The Land of Little Rain.*

11. Lee, *Water Resources of a Part of Owens Valley,* 77.

12. DeDecker, "Owens Valley, Then and Now," 7–15.

13. Watson and Storie, *Soil Survey of the Bishop Area, California,* 68.

14. Ibid., 67–71.

15. Steward, "Irrigation Without Agriculture," 149.

16. Steward, *Basin-Plateau Aboriginal Sociopolitical Groups,* 33, 52.

17. Bettinger, "Aboriginal Human Ecology in Owens Valley," 5.

18. Steward, *Basin-Plateau Aboriginal Sociopolitical Groups,* 51–52.

19. Worster, *Rivers of Empire,* 32.

20. Steward, "Irrigation Without Agriculture," 150.

21. Lawton et al., "Agriculture Among the Paiute of Owens Valley," 13–50; Walton, *Western Times and Water Wars,* 15.

22. Worster, *Rivers of Empire,* 32; Lawton et al., "Agriculture Among the Paiute of Owens Valley," 15; Steward, *Ethnography of the Owens Valley Paiute,* 249.

23. Steward, *Ethnography of the Owens Valley Paiute,* 247.

24. Ibid.

25. Lawton et al., "Agriculture Among the Paiute of Owens Valley," 20; Steward, *Ethnography of the Owens Valley Paiute,* 242–46; Davidson, *Expedition,* 47.

26. Steward, *Ethnography of the Owens Valley Paiute,* 239.

27. Lawton et al., "Agriculture Among the Paiute of Owens Valley," 42.

28. Steward, *Ethnography of the Owens Valley Paiute,* 248.

29. Ibid., 249.

30. Lawton et al., "Agriculture Among the Paiute of Owens Valley," 15.

31. Chalfant, *The Story of Inyo,* 44; Steward, *Ethnography of the Owens Valley Paiute,* 237; Davidson, *Expedition,* 29, 46.

32. Steward, *Basin Plateau Aboriginal Sociopolitical Groups,* 51–52, 58 (map).

33. Ibid.

Chapter 3. Early Views of the Land

1. Wagner, *Leonard's Narrative,* 231.

2. Ibid., 232.

3. Ibid., 231.

4. Davidson, *Expedition,* 9; Chalfant, *The Story of Inyo,* 98.

5. Chalfant, *The Story of Inyo,* 98.

6. Kern used the name "Owen's River" in his journal on December 16, 1845, presumably before Frémont had given the valley its name. Wilke and

Lawton therefore believe that Kern should be credited with naming the region; see Davidson, *Expedition,* 36–37.

7. Simpson, "Journal of Mr. Edward M. Kern," 482.

8. Ibid.

9. Preston, *Vanishing Landscapes,* 76.

10. U.S., Commissioner of the General Land Office, *Instructions to the Surveyors General of Public Lands.*

11. Von Schmidt, Field Notes, vol. 105-14, July 15, 1855, pp. 347–48.

12. Chalfant, *The Story of Inyo,* 121.

13. Von Schmidt, Field Notes, vol. 259-8, August 4, 1856, p. 9.

14. Von Schmidt, Field Notes, vol. 114-8, November 6, 1856, p. 173.

15. Preston, *Vanishing Landscapes,* 77.

16. Davidson, *Expedition,* 14–15. An order by the federal government in February 1859 withdrew T13S R35E from settlement pending a decision on whether to locate an Indian reservation there. Senator Milton Latham introduced the idea as a bill in Congress in 1862, but the order of withdrawal was revoked in 1864. See Chalfant, *The Story of Inyo,* 124.

17. Davidson, *Expedition,* 16–31. Davidson's original manuscript is housed in the Navy and Old Army Branch, Military Archives Division, National Archives and Records Service, Washington, D.C.

18. See Wilke and Lawton, "The Account of a Newspaper Correspondent," in Davidson, *Expedition,* 32–35.

19. Ibid., 25, 26, 27. Most of the names that Davidson gave to streams and other topographic features were later superseded. For example, Davidson's Winn Valley became Round Valley. The names used by Davidson have been translated here into current usage.

20. Ibid., 20.

21. Ibid., 27. The Bishop Creek and Oak Creek stands were corroborated by Von Schmidt in his field notes. Von Schmidt may have discontinued his survey because of rugged topography before reaching the pines on Big Pine Creek.

22. Ibid., 26, 27.

23. Farquar, *Up and Down California in 1860–1864: The Journal of William H. Brewer,* 536, 538.

Chapter 4. Thoroughfare to the Mines

1. Chalfant, *The Story of Inyo,* 127.

2. McGrath, *Gunfighters, Highwaymen and Vigilantes,* 1.

3. Chalfant, *The Story of Inyo,* 142.

4. Ibid.

5. McGrath, *Gunfighters, Highwaymen and Vigilantes,* 17.

6. Quoted in Lawton et al., "Agriculture Among the Paiute of Owens Valley," 30.

7. Busby et al., *A Cultural Resource Overview,* 54.

8. Davidson, *Expedition,* 31.

9. *Official Records of the Union and Confederate Armies,* pt. 1, vol. 50, p. 936.

10. Ibid., 49.

11. Ibid., 50.

12. Ibid., 146.

13. McGrath, *Gunfighters, Highwaymen and Vigilantes,* 32.

14. Quoted in McGrath, *Gunfighters, Highwaymen and Vigilantes,* 32.

15. Ibid., 50.

16. Brewer, *Up and Down California in 1860–1864,* 537.

17. Busby et al., *A Cultural Resource Overview,* 47; Chalfant, *The Story of Inyo,* 240.

18. Ibid. A small portion of southern Mono County was also taken by Inyo, originally marking the new county's northern boundary at Big Pine Creek. Residents of Round Valley and Bishop, however, felt inconvenienced by having to travel to Bridgeport for county business. Four years later, in 1870, the present northern boundary of Inyo County was established when the southern portion of Mono County was purchased for $12,000. A final boundary change occurred in 1872, when the southern line of the county, which extended in a northeasterly direction when the land was first taken from Tulare County, was changed to an east–west line.

19. See, for example, Patterson et al., *Nevada's Northeast Frontier.*

Chapter 5. The Process of Pioneer Settlement, 1860–1880

1. Bowman, *The Pioneer Fringe,* 6.

2. Gates, *History of Public Land Law Development,* 238.

3. Sauder and Sauder, "The Morrill Act's Influence on Public Land Disposal After 1870," 45.

4. Gates, *History of Public Land Law Development,* 395.

5. McIntosh, "Use and Abuse of the Timber Culture Act," 347.

6. Gates, *History of Public Land Law Development,* 637–38.

7. Robinson, *Land in California,* 190.

8. Gates, "Public Land Disposal in California," 161.

9. Nash, "The California State Land Office," 349.

10. Robinson, *Land in California,* 192–93; Gates, "Public Land Disposal in California," 165.

11. Gates, *History of Public Land Law Development,* 327.

12. Von Schmidt, Field Notes, vol. 116-4, September 29, 1856, p. 89; vol. 116-8, October 2, 1856, p. 309.

13. "A Stupendous Grab," *Inyo Independent,* October 4, 1873, p. 2. The basis for the Owens Valley's swampland designation was a ruling of the commis-

sioner of the General Land Office that "lands too wet for cultivation at the usual seeding time, though requiring irrigation later in the season, may be segregated as swamp and overflowed"; ibid.

14. "Something About the Grab," *Inyo Independent*, October 25, 1873, p. 2.

15. "Land Monopoly," *Inyo Independent*, November 15, 1873, p. 2.

16. Chalfant, *The Story of Inyo*, 317.

17. Since only final preemption entries were recorded in the land office tract books, the number of preemptors who abandoned their claims in the Owens Valley cannot be determined. There are only twenty abandoned preemption files among the land entry papers of the Independence District Land Office housed in the National Archives.

18. Compiled from the Independence District Land Office tract books, master title plats, and historical indexes. In Gage County, Nebraska, in a region where the homestead principle was first applied, the proportion of successful homestead entries averaged 60 percent. See Okada, *Public Lands and Pioneer Farmers*, 43.

19. Jackson, "Mormon Perception and Settlement."

20. An excellent discussion of the land entry and land patent process, as well as the nature of the data sources, is found in McIntosh, "Patterns from Land Alienation Maps." The term *patented entries* as used here includes entries that had been patented by the date indicated as well as those that had been entered by that date and which ultimately would be carried to patent.

21. Preston, *Vanishing Landscapes*, 115.

22. *Inyo Independent*, May 1, 1875, p. 2.

23. "Owens River Valley," *Inyo Independent*, January 6, 1877, p. 2.

24. Von Schmidt, Field Notes, vol. 116-11, September 28, 1856, p. 369.

25. Ibid., vol. 116-8, October 2, 1856, p. 309.

26. Whitney, "The Owen's Valley Earthquake," 36.

27. Von Schmidt, Field Notes, vol. 114-8, November 6, 1856, p. 173; vol. 114-9, November 3, 1856, p. 205; vol. 114-10, October 29, 1856, p. 226; vol. 114-16, October 26, 1856, p. 378; vol. 114-18, October 23, 1856, p. 430; and vol. 115-17, October 18, 1856, p. 423.

28. "Inyo County," *Inyo Independent*, July 13, 1870, p. 2.

29. "The Story of an Irrigated Farm," *Sierra Magazine* 2 (January 1909).

30. "Owens River Valley," *Inyo Independent*, January 6, 1877, p. 2.

31. Ibid.

32. Similarly, nearly twenty years later, in the early 1890s, pioneer settlers in the Greybull River country of Wyoming used dense clusters of sagebrush growing to heights of ten to twelve feet as an indicator of superior land; Bowen, "Mormon Land Selection in the Greybull Valley," 42–43.

33. Watson and Storie, *Soil Survey of the Bishop Area, California*, 72.

Chapter 6. Establishing A Working Economy, 1860–1880

1. Several months prior to the initiation of the census, the current northern boundary of Inyo County was established, and since nearly all of the inhabited and cultivated areas of Inyo County have always been located in the Owens Valley, the agricultural statistics for the valley are essentially those for the entire county. Chalfant mistakenly believed that the region north of Big Pine Creek had not been included in the 1870 federal census of Inyo County; *The Story of Inyo,* 44.

2. Hart, "The Middle West," 267.

3. Wentworth, *America's Sheep Trails,* 193.

4. Compiled from U.S., Ninth Census, 1870, Manuscript Census of Inyo County, Schedule 3.

5. The 1870 federal manuscript census schedules do not subdivide the valley into its respective communities. Regional subdivisions were derived by correlating names listed on the manuscript census rolls with land ownership data contained in the Independence District Land Office tract books.

6. Walton, *Western Times and Water Wars,* 25–29.

7. Busby et al., *A Cultural Resource Overview,* 57; Earl, *The Enchanted Valley,* 148; Walton, *Western Times and Water Wars,* 39–40.

8. Walton, *Western Times and Water Wars,* 40.

9. Ibid., 39.

10. "A New Indian Reservation," *Inyo Independent,* January 4, 1873, p. 2.

11. Ibid.

12. Walton, *Western Times and Water Wars,* 24.

13. "Indian Homesteads," *Inyo Independent,* March 27, 1875, p. 2. The Indians living in the northern part of the valley in particular availed themselves of the privilege of entering lands under this federal provision. In a single week in late September, for example, it was reported that "some fifteen or sixteen [Indians] filed on Homesteads"; "Indian Homesteads," *Inyo Independent,* October 2, 1875, p. 3.

14. Earl, *The Enchanted Valley,* 35, 39, 41, 42, 113.

15. Chalfant, *The Story of Inyo,* 213–14.

16. "The No-Fence Law," *Inyo Independent,* July 22, 1871, p. 2.

17. Chalfant, *The Story of Inyo,* 318.

18. Earl, *The Enchanted Valley,* 46–47.

19. "Among the Farms," *Inyo Independent,* July 28, 1877, p. 3.

20. Earl, *The Enchanted Valley,* 42.

21. L. T. Larsen, "Hardwood Planting in Owens Valley, California," July 1, 1914, p. 9.

22. "The No-Fence Petition," *Inyo Independent,* March 2, 1872, p. 2.

23. "The No-Fence Question," *Inyo Independent,* February 3, 1872. p. 2.

24. "The No-Fence Petition," *Inyo Independent,* March 2, 1872, p. 2.

25. "No-Fence Law Approved by State Legislature for Inyo County," *Inyo Independent,* June 1, 1872, p. 1.

26. "For a Change in the Fence Law," *Inyo Independent,* November 27, 1875, p. 2.

27. "Bishop and Round Valley," *Inyo Independent,* March 4, 1876, p. 3.

28. "Among the Farms," *Inyo Independent,* July 28, 1877, p. 3.

29. Ibid.

30. "Inyo Statistics," *Inyo Independent,* July 6, 1878, p. 2.

31. U.S., Tenth Census, 1880, Manuscript Census of Inyo County, Schedule 2.

32. "Fruit Destroyed," *Inyo Independent,* August 22, 1870, p. 3.

33. Chalfant, *The Story of Inyo,* 330.

34. Busby et al., *A Cultural Resource Overview,* 66.

35. "The Need of a Railroad to Our Farming Interest," *Inyo Independent,* May 5, 1877, p. 2.

36. "The Bishop Creek Country," *Inyo Independent,* June 5, 1880, p. 2.

37. "Twenty Years Ago – Then and Now," *Inyo Register,* April 4, 1885, p. 2.

38. "Among the Farms," *Inyo Independent,* July 28, 1877, p. 3.

39. "About Crops," *Inyo Independent,* September 30, 1871, p. 3.

Chapter 7. Economic Organization and Environmental Alteration, 1880–1905

1. "Letter from Round Valley," *Inyo Independent,* November 3, 1883, p. 2.

2. *Inyo Register,* April 16, 1885, p. 2.

3. *Inyo Register,* April 10, 1890, p. 3.

4. "A Visitor's Estimate," *Inyo Register,* June 6, 1893, p. 3, reprinted from the *Los Angeles Times.*

5. "Owens River Valley," *Inyo Independent,* January 6, 1877, p. 3.

6. Ibid.

7. Ganoe, "The Origin of a National Reclamation Policy," 34.

8. "The Desert Land Act," *Inyo Independent,* August 25, 1877, p. 2.

9. "Land Entries," *Inyo Independent,* April 14, 1877, p. 3.

10. Ganoe, "The Desert Land Act in Operation," 143; Kollmorgen, "The Woodsman's Assaults on the Domain of the Cattleman," 222.

11. Ganoe, "The Desert Land Act in Operation," 144.

12. Ibid., 151.

13. "The Lone Pine Irrigating Canal," *Inyo Independent,* April 14, 1877, p. 3.

14. "The Bishop Creek Country," *Inyo Independent,* June 5, 1880, p. 2.

15. "About Ditches," *Inyo Register,* April 4, 1885, p. 3.

16. Gates, "The Homestead Law in an Incongruous Land System," 656.

17. "Land and Water," *Inyo Register,* December 16, 1886, p. 2.

18. "Two Kinds of Speculation," *Inyo Independent,* July 27, 1889, p. 3.

19. "Lands for the Landless," *Inyo Independent,* March 13, 1886, p. 3; "Land Abuses," *Inyo Independent,* April 9, 1887, p. 3.

20. In 1891 Congress amended the Desert Land Act to impose more demanding requirements in an attempt to remedy land abuses. Desert land locators were now required to file a map indicating how they intended to irrigate their land and prove that the water supply was adequate for the purpose. They were also required at the time of final entry to show that they had expended funds to secure water for land improvements. Ganoe, "The Desert Land Act in Operation," 156.

21. "The New Era of Ditcher," *Inyo Independent,* November 12, 1887, p. 2.

22. "Two Kinds of Speculation," *Inyo Independent,* July 27, 1889, p. 3.

23. "Proving Up," *Inyo Independent,* February 20, 1891, p. 3.

24. "Unpleasant Facts," *Inyo Independent,* February 20, 1891, p. 3.

25. "As It Appears," *Inyo Register,* January 28, 1892, p. 2.

26. "Inyo County, California," *Inyo Register,* April 18, 1901, p. 1.

27. Compiled from the Independence District Land Office tract books, master title plats, and historical indexes.

28. Gates, *History of Public Land Law Development,* 642.

29. *Inyo Register,* March 18, 1886, p. 2.

30. "Notes by the Way," *Inyo Independent,* August 16, 1884, p. 2.

31. *Inyo Independent,* July 30, 1887, p. 2.

32. *Inyo Independent,* July 11, 1890, p. 2.

33. "As It Appears," *Inyo Register,* January 28, 1892, p. 2.

34. Chalfant, *The Story of Inyo,* 327.

35. "Unpleasant Facts," *Inyo Independent,* February 20, 1891, p. 3.

36. "What Farmers Say," *Inyo Independent,* March 6, 1891, p. 3.

37. Ibid.

38. Quoted in "A Truthful Statement," *Inyo Register,* January 6, 1898, p. 2.

39. The three census townships that included settlement in the Owens Valley recorded an increase of 560 persons between 1890 and 1900. "Some Census Figures," *Inyo Register,* January 10, 1901. p. 3.

40. *Inyo Independent,* May 15, 1903, p. 2.

41. "Notes by the Way," *Inyo Independent,* August 16, 1884, p. 2.

42. "Notes by the Way," *Inyo Independent,* August 9, 1884, p. 2.

43. Ibid.; "The Sanger Ranch," *Inyo Independent,* August 13, 1887, p. 3. The grubbing machine typically consisted of two pieces of iron rail bolted together, each sixteen feet long, with the flange of one piece lapping the head of the other. With eyebolts at either end, the implement was hitched to a team of four horses and dragged over the ground. Two trips were required across each strip, one to pull the brush from the ground and the other to gather it into bunches for burning. Using this method, a four-horse team could clear about fifteen acres in a day.

44. "Continued Calls for Information," *Inyo Independent,* August 7, 1886, p. 2.

45. Mulholland, *Inyo County*, 13.

46. "North End of Owens Valley," *Inyo Independent*, September 29, 1893, p. 3.

47. *Inyo Register*, November 16, 1893, p. 3.

48. Mulholland, *Inyo County*, 11.

49. "Round Valley Jottings," *Inyo Register*, May 15, 1890, p. 2.

50. "Round Valley Jottings," *Inyo Register*, December 22, 1898, p. 3.

51. "Vineyards," *Inyo Register*, May 1, 1890, p. 3.

52. "A Rapidly Growing Section," *Inyo Register*, March 7, 1895, p. 3.

53. Mulholland, *Inyo County*, 15–16.

54. *Inyo Register*, February 17, 1898, p. 3.

55. "March of Progress," *Inyo Independent*, May 19, 1899, p. 3.

56. Ibid.

57. "North End of Owens Valley," *Inyo Independent*, September 29, 1893, p. 3.

58. "March of Progress," *Inyo Independent*, May 19, 1899, p. 3.

59. Ibid.

60. Mulholland, *Inyo County*, 17.

61. Ibid., 22.

62. Clausen, "Report of the Owens Valley," 67.

63. "Round Valley Jottings," *Inyo Register*, December 22, 1898, p. 3. One piece of ground in Round Valley was reportedly supporting its seventeenth successive crop in 1884; "Notes By the Way," *Inyo Independent*, August 16, 1884, p. 2.

64. "Railroad," *Inyo Independent*, February 16, 1900, p. 3.

65. Cotton, "Agricultural Conditions of Inyo County," 4.

66. "Too Much Water," *Inyo Independent*, March 5, 1887, p. 2.

67. Ibid.

68. "Big Pine Items," *Inyo Register*, May 19, 1904, p. 3.

69. Lee, *Geology and Water Resources*, 23. An acre-foot of water is equivalent to 325,000 gallons.

70. Clausen, "Report of the Owens Valley," 17.

71. *Inyo Independent*, November 2, 1894, p. 3.

72. *Inyo Register*, September 8, 1898, p. 3.

73. *Inyo Independent*, August 4, 1899, p. 3.

74. "Is Owens Lake Near Its End?" *Inyo Register*, August 4, 1904, p. 1; Clausen, "Report of the Owens Valley," 21. The maximum depth of the lake had diminished from 51 feet to 36 feet.

75. "Is Owens Lake Near Its End?" *Inyo Register*, August 4, 1904, p. 1.

76. Lee, *Geology and Water Resources*, 21.

77. "Is Owens Lake Near Its End?" *Inyo Register*, August 4, 1904, p. 1.

78. Quoted in "A Review of the County," *Inyo Register*, January 3, 1901, p. 3.

79. "Inyo's Present and Future," *Inyo Register*, January 1, 1903, p. 2.

80. "Inyo County Improvements," *Inyo Independent,* January 22, 1904, p. 3, reprinted from the *Los Angeles Times.*

81. Baugh, "Land Use Changes in the Bishop Area of Owens Valley," 23.

82. Powell, *Report on the Lands of the Arid Region of the United States,* 25–45. Major Powell's arid lands survey did not reach into California. Instead, it focused on the plateau country of Utah, western Colorado, and northern Arizona, with the most intensive work in Utah. His recommended adjustments to the land disposal system were therefore largely based on information obtained in Utah.

Chapter 8. Federal Reclamation and Municipal Manipulation, 1900–1913

1. Editorial, *Inyo Register,* January 12, 1905, p. 2.

2. *Inyo Register,* July 3, 1902, p. 2; *Inyo Register,* July 31, 1902, p. 2.

3. Lippincott, "The Reclamation Service in California," 164.

4. "Inyo County, California," *Inyo Register,* April 18, 1901, p. 1.

5. "Waste of Water," *Inyo Independent,* February 27, 1903, p. 3.

6. *Inyo Independent,* September 16, 1892, p. 3; *Inyo Independent,* July 7, 1893, p. 3; *Inyo Register,* July 13, 1893, p. 3.

7. U.S. Congress, Senate, *Report of the Special Committee . . . on the Irrigation and Reclamation of Arid Lands,* 2:329–30.

8. Ibid., 331.

9. According to the *Inyo Register,* the *Los Angeles Express* reported that Eaton had conceived his plan as early as 1890. This would place the inception of the idea soon after the Senate hearings on the irrigation and reclamation of arid lands. "Los Angeles Plots Destruction," *Inyo Register,* August 3, 1905, p. 3.

10. Kahrl, *Water and Power,* 48.

11. "Lands Withdrawn," *Inyo Register,* September 3, 1903, p. 2; Chalfant, *The Story of Inyo,* 339.

12. U. S. Geological Survey, *Second Annual Report of the Reclamation Service, 1902–3,* 27; "Lands Withdrawn," *Inyo Register,* September 3, 1903, p. 2; "Public Land Withdrawals," *Inyo Register,* February 9, 1905, p. 3.

13. "The Irrigation Project," *Inyo Register,* November 26, 1903, p. 3.

14. Lippincott, "The Reclamation Service in California," p. 164.

15. Clausen, "Report of the Owens Valley," 15–18.

16. Ibid., 17.

17. Kahrl, *Water and Power,* 54.

18. Clausen, "Report of the Owens Valley," 15.

19. Ibid., 70–71.

20. Chalfant, *The Story of Inyo,* 340–41; Kahrl, *Water and Power,* 57.

21. Los Angeles, Aqueduct Investigation Board, "Report," 7.

22. Kahrl, *Water and Power,* 64.

23. Ibid., 42.

24. "Something Doing," *Inyo Register*, April 13, 1905, p. 2.
25. "A Large Realty Transaction," *Inyo Register*, June 15, 1905, p. 3.
26. Hoffman, *Vision or Villainy*, 79–81.
27. "More Sales Recorded," *Inyo Independent*, June 30, 1905, p. 3.
28. Los Angeles, Aqueduct Investigation Board, "Report," 8.
29. *Inyo Register*, August 3, 1905, p. 3.
30. "In Unity Is the Only Hope," *Inyo Register*, August 3, 1905, p. 3.
31. *Inyo Independent*, August 4, 1905, p. 3.
32. "Best We Could Do," *Inyo Independent*, August 18, 1905, p. 3.
33. See, for example, Hoffman, *Vision or Villainy*, 77; Kahrl, *Water and Power*, 64.
34. Quoted in Hoffman, *Vision or Villainy*, 53.
35. *U.S. Statutes at Large*, 55th Cong., 1st Sess., vol. 34, pt. 1, chap. 3926, p. 803.
36. Miller, "Los Angeles and the Owens River Aqueduct," 78.
37. Ibid., 125–26.
38. Los Angeles, Aqueduct Investigation Board, "Report," 7.
39. "Valley Wins All Contested Points," *Inyo Register*, May 11, 1911, p. 1.
40. "Los Angeles Wins All," *Inyo Register*, January 12, 1912, p. 2.
41. *Inyo Register*, September 9, 1909, p. 2.
42. "Irrigation District Proposed," *Inyo Register*, July 8, 1909, p. 2. The *Register* reminded readers that it had recommended the formation of an irrigation district as early as 1891 but that the idea had not been pursued by valley residents.
43. "For Irrigation District," *Inyo Register*, May 19, 1910, p. 1.
44. Historical index to T6S R31E, Mount Diablo Meridian.
45. Chalfant, *The Story of Inyo*, 364; Kahrl, *Water And Power*, 211; Hoffman, *Vision or Villainy*, 138–40; Nadeau, *The Water Seekers*, 38.
46. Chalfant, *The Story of Inyo*, 364.
47. "A Theft in Water," *Inyo Magazine*, November 1, 1908.
48. Los Angeles, Aqueduct Investigation Board, "Report," 9.
49. *Inyo Register*, March 5, 1908, p. 2.
50. *Inyo Independent*, May 21, 1909, p. 3.
51. Compiled from the Independence District Land Office tract books.
52. Kahrl, *Water and Power*, 207.
53. Chalfant, *The Story of Inyo*, 368.
54. Kahrl, *Water and Power*, 216.
55. Kahrl, "The Politics of California Water," 99.
56. Miller, "Los Angeles and the Owens River Aqueduct," 229.
57. Kahrl, *Water and Power*, 137–38.
58. Ibid., 131.
59. Hoffman, *Vision or Villainy*, 123.
60. Ibid., 124–25.
61. Ibid., 141.

62. Los Angeles, Board of Public Service Commissioners, *Owens Valley,* 10.

63. Blanchard, "Redeeming the West," 209.

64. Worster, *Rivers of Empire,* 278.

Chapter 9. The Intensification of Agriculture, 1900–1925

1. Hoffman, *Vision or Villainy,* 174.

2. "An Inyo County Farm," *Inyo Independent,* November 8, 1907, p. 3.

3. "Will Plant Orchards," *Inyo Independent,* January 28, 1910, p. 3.

4. Owens Valley Improvement Company, "Fortunes in Apples in Owens Valley, Inyo County, California," 8.

5. "Many New Settlers Buying Homes Near Independence," *Inyo Independent,* January 27, 1911, p. 1.

6. Edwin Schallert, "The Valley of Flowing Waters," *West Coast Magazine* 11 (November 1911): 145.

7. "Owens Valley Dairying," *Inyo Independent,* October 6, 1916, p. 1.

8. "The Possibilities of Owens River Valley," *Inyo Independent,* April 8, 1922, p. 2.

9. "Make Dairying Pay," *Inyo Independent,* January 28, 1922, p. 1.

10. "Sugar Beet Industry," *Inyo Independent,* October 13, 1916, p. 1.

11. "A Sugar Beet Factory," *Inyo Independent,* October 27, 1916, p. 1.

12. "Beets to be Planted," *Inyo Independent,* November 24, 1916, p. 1.

13. California Development Board, "Agricultural and Industrial Survey of Inyo County, California," San Francisco, September 1917, 1: Sugar Beets.

14. U.S., *Fourteenth Census, 1920,* vol. 6, pt. 3, table 4, 358.

15. California Development Board, "Agricultural and Industrial Survey," 1: Apples.

16. "Industrial Survey of County," *Inyo Independent,* June 8, 1917, p. 1.

17. California Development Board, "Agricultural and Industrial Survey," 1: Pears.

18. "Will Develop Valley," *Inyo Independent,* November 27, 1920, p. 1.

19. Ibid.

20. "Pear Growers' Association Will Solve Future of Valley," *Inyo Independent,* December 11, 1920, p. 1, reprinted from the *Owens Valley Herald.*

21. "Big Acreage Being Planted," *Inyo Independent,* March 26, 1921, p. 2.

22. "Owens Valley to Become Famous for Pears," *Inyo Independent,* March 5, 1921, p. 3, reprinted from the *Owens Valley Herald.*

23. "Inyo County's Prosperity Shown During Past Year," *Inyo Independent,* January 7, 1922, p. 3, reprinted from the *Los Angeles Times.*

24. "Fine for Fruit Crop," *Inyo Independent,* February 4, 1922, p. 1.

25. U.S., *Fourteenth Census, 1920,* vol. 6, pt. 3, table 4, 358; U.S., *Census of Agriculture: 1925,* table 4, 474.

26. "Conserving Water," *Inyo Independent,* April 3, 1920, p. 1.

27. "Outlook in Owens Valley Bright for Coming Year," *Inyo Independent*, February 4, 1922, p. 2.

28. "Drainage Problems to Have Attention," *Inyo Register*, March 1, 1917, p. 1.

29. Conkling, "Report on Owens Valley Project," 26; McCarthy and Clausen, "Engineers' Report on the Owens Valley Irrigation District," 17.

30. Watson and Storie, *Soil Survey of the Bishop Area*, 93.

31. California Development Board, "Agricultural and Industrial Survey," 2: Irrigation and Drainage.

32. Kahrl, *Water and Power*, 256.

33. "Drainage of Lands Urged by Experts," *Inyo Register*, March 22, 1917, p. 1.

34. *Inyo Register*, February 27, 1919, p. 2.

35. "Drainage of Lands Urged by Experts," *Inyo Register*, March 22, 1917, p. 1.

Chapter 10. The Lost Frontier

1. "1922," *Inyo Independent*, January 7, 1922, p. 2.

2. *Inyo Register*, January 7, 1925, p. 1.

3. McClure, *Letter of Transmittal*, 19.

4. "Valley Welfare in Organizing District," *Inyo Register*, February 19, 1920, p. 1.

5. "District Needed for Good of Valley," *Inyo Register*, February 26, 1920, p. 1.

6. "District Needed by Irrigation Interests," *Inyo Register*, March 4, 1920, p. 1.

7. "What Has Become of the Irrigation District," *Inyo Register*, August 18, 1921, p. 2.

8. *Inyo Register*, August 25, 1921, p. 1.

9. "Irrigation District Petition Will Be Presented Saturday Night," *Inyo Register*, May 25, 1922, p. 1.

10. "Almost Unanimous Vote for Irrigation District: 596 For, 27 Against," *Inyo Register*, December 28, 1922, p. 1.

11. McCarthy and Clausen, "Engineers' Report on the Owens Valley Irrigation District," 1.

12. Ibid., 17–18.

13. "Los Angeles Decides to Buy Big Pine Canal Water," *Inyo Independent*, November 3, 1923, p. 2, reprinted from the *Owens Valley Herald*. Los Angeles paid $1,100,000 for the Big Pine Canal lands.

14. Ibid.

15. "Must Condemn Water Rights," *Inyo Independent*, February 23, 1924, p. 3, reprinted from the *Owens Valley Herald*.

16. "Condemnation Suits Brought," *Inyo Independent*, March 8, 1924, p. 4, reprinted from the *Owens Valley Herald*.

17. Nadeau, "The Water War," 103.

18. McClure, *Letter of Transmittal*, 11.

19. Walton, "Picnic at Alabama Gates," 203.

20. McClure, *Letter of Transmittal*, 28.

21. Walton, *Western Times and Water Wars*, 205.

22. U.S. Congress, Senate, *Survey of Conditions of the Indians in the United States*, pt. 29, p. 15260.

23. Walton, "Picnic at Alabama Gates," 203.

24. "Farm Bureau Committee's Report on Water Situation," *Inyo Independent*, July 12, 1924, p. 1.

25. Nadeau, "The Water War," 103.

26. Kahrl, "The Politics of California Water," 110.

27. Chalfant, *The Story of Inyo*, 337.

28. Miller, "Los Angeles and the Owens River Aqueduct," 161.

29. City of Los Angeles, Dept. of Water and Power, "Land and Water Right Purchases," 2.

30. Samuel Adams to President William Howard Taft, March 11, 1912, quoted in Kahrl, *Water and Power*, 353.

31. Ibid.

32. Walton, *Western Times and Water Wars*, 207.

33. U.S. Congress, Senate, *Survey of the Conditions of the Indians*, 15235.

34. Walton, *Western Times and Water Wars*, 207–8; Kahrl, *Water and Power*, 357.

35. Kahrl, *Water and Power*, 356.

36. Ibid., 358; Walton, *Western Times and Water Wars*, 209.

37. Kahrl, *Water and Power*, 358.

38. Ibid., 359. Nancy Peterson Walter contends, however, that the agreement overlooked water rights of the Paiute in connection with the Winters Doctrine and hence that the U.S. government failed to carry out its trust responsibility in the Los Angeles exchange; "The Land Exchange Act of 1937," 379.

39. Baugh, "Land Use Changes in the Bishop Area of Owens Valley," 28.

40. Ibid., 26.

41. Ibid., 31.

42. Nadeau, "The Water War," 107.

43. Kahrl, *Water and Power*, 440.

Chapter 11. Water Transfers and Rural Preservation

1. Thomas J. Knudson, "Dry Cities of West Buy Up Farm Water Rights," *New York Times*, February 10, 1987, p. A-1.

2. Saliba and Bush, *Water Markets in Theory and Practice*, 245–46.

3. Saliba, "Do Water Markets 'Work'?" 1117.

4. Brown et al., "Water Reallocation," 233; Anderson, "Discussion," 429; Knudson, "Dry Cities of West," 1.

5. Brown et al., "Water Reallocation," 238; Wahl and Osterhoudt, "Voluntary Transfers of Water in the West," 120.

6. Roderick, "Las Vegas' Thirst for Water Upsets Many in Arid West," *Los Angeles Times,* May 6, 1991, p. A-1.

7. "Battle Lines Drawn in Sand as Las Vegas Covets Water," *New York Times,* April 23, 1991, p. A-1; Virginia Ellis, "Urban Water Bill OKd by Assembly," *Los Angeles Times,* July 3, 1991. p. A-3.

8. DeDecker, "Owens Valley, Then and Now," 13.

9. EIP Associates, *Water From the Owens Valley,* 1:10–30.

10. Kevin Roderick, "Owens Lake's Dust: Airborne Health Hazard," *Los Angeles Times,* April 2, 1989, pt. 1, p. 1.

11. EIP Associates, *Water From the Owens Valley,* 1:10–46.

12. Ibid.

13. Busby et al., *A Cultural Resource Overview,* 97.

14. Frederick M. Muir, "DWP Signs Deal With Inyo County on Water Rights," *Los Angeles Times,* October 16, 1991, p. B-1.

15. Kahrl, *Water and Power,* 350

16. "A Higher Responsibility," *Los Angeles Times,* July 26, 1989, pt. 2, p. 6.

17. Robert A. Jones, "Still Waters on Battlefields Run Deep," *Los Angeles Times,* August 15, 1989, pt. 1, p. 3.

18. See Sauder, "Mountains and Lowlands," (forthcoming).

19. Wahl and Osterhoudt, "Voluntary Transfers of Water in the West," 113.

20. Frederick, *Water for Western Agriculture,* 113–14.

21. Saliba and Bush, *Water Markets in Theory and Practice,* 80.

22. Ingram, "Water Rights in the Western States," 141.

23. Wilkinson, "Western Water Law in Transition," 40.

24. Weatherford and Shupe, "Reallocating Water in the West," 67.

25. Wahl and Osterhoudt, "Voluntary Transfers of Water in the West," 116–17.

26. "Water: This Time It's Serious," *Los Angeles Times,* December 24, 1989, p. M-6.

27. A detailed discussion of the community value of water is found in Brown and Ingram, *Water and Poverty in the Southwest.*

28. Bruce Babbitt, "Urbanized West Needs New Water Laws," *Los Angeles Times,* August 9, 1988, pt. 2, p. 7.

29. Udall, "Just Add Water Marketing," 41.

30. Babbitt, "Urbanized West Needs New Water Laws," 7.

Bibliography

Anderson, D. H. "Are Government Officials Playing Fair? Some Facts About the Owens River Valley Condition." *Irrigation Age* 24 (December 1908): 40–41.

———. "Oppressive Use of Power." *Irrigation Age* 24 (January 1909): 71–72.

———. "Why Was the Owens River Project Abandoned?" *Irrigation Age* 24 (February 1909): 103–5.

Anderson, Raymond L. "Discussion." In *Water Scarcity: Impacts on Western Agriculture,* edited by Earnest A. Engelburt, 427–30. Berkeley: University of California Press, 1984.

Arrington, Leonard J. *Great Basin Kingdom: An Economic History of the Latter-day Saints, 1830–1900.* Cambridge, Mass.: Harvard University Press, 1958.

Austin, Mary. *The Land of Little Rain.* Boston: Houghton, Mifflin, 1903.

Baugh, Ruth E. "Land Use Changes in the Bishop Area of Owens Valley, California." *Economic Geography* 13 (January 1937): 17–34.

Bettinger, Robert L. "Aboriginal Human Ecology in Owens Valley: Prehistoric Change in the Great Basin." *American Antiquity* 42 (January 1977): 3–17.

Blackwelder, Eliot. "Pleistocene Lakes and Drainage in the Mojave Region, Southern California." In *Geology of Southern California,* edited by Richard H. Jahns, 35–40. California Division of Mines Bulletin 170. Sacramento: Department of Natural Resources, 1954.

Blanchard, C. J. "Redeeming the West: What the Great Klamath Project Means in Government Reclamation." *Sunset* 17 (September 1906): 207–14.

Bowen, Marshall. "Mormon Land Selection in the Greybull Valley, Wyoming: Rationale and Implications." *Ecumene* 8 (March 1976): 40–49.

———. "Elko County's Dry Farming Experimental Station." *Northeastern Nevada Historical Society Quarterly* 2 (Spring 1979): 34–51.

———. "Turnover of Pioneers and Property in a Marginal Nevada Farming Community." *Yearbook of the Association of Pacific Coast Geographers* 42 (1980): 45–57.

———. "Jackrabbit Invasion of a Nevada Agricultural Community." *Ecumene* 12 (October 1980): 6–16.

———. "Dryland Homesteading on Tobar Flat." *Northeastern Nevada Historical Society Quarterly* 4 (Fall 1981): 118–40.

———. "Pioneer Success and Failure in a Northeastern Nevada Valley." *Pioneer America Society Transactions* 9 (1986): 23–29.

———. "The Heritage of an Empty Land: Independence Valley, Nevada." *Journal of Cultural Geography* 6 (Spring/Summer 1986): 67–79.

———. "The Desert Homestead as a Non-Farm Residence." *Nevada Historical Society Quarterly* 31 (Fall 1988): 198–211.

———. "A Backward Step: From Irrigation to Dry Farming in the Nevada Desert." *Agricultural History* 63 (Spring 1989): 231–42.

Bowman, Isaiah. *The Pioneer Fringe*. New York: American Geographical Society, 1931.

Brewer, William H. *Up and Down California in 1860–1864: The Journal of William H. Brewer*. Edited by Francis P. Farquar. Berkeley: University of California Press, 1974.

Brown, F. Lee, and Helen M. Ingram. *Water and Poverty in the Southwest*. Tucson: University of Arizona Press, 1987.

Brown, F. Lee, J. Tysseling McDonald, and C. Dumars. "Water Reallocation, Market Proficiency, and Conflicting Social Values." In *Water and Agriculture in the Western U.S. Conservation, Reallocation and Markets*, edited by Gary Weatherford, 191–255. Boulder: Westview Press, 1982.

Busby, Colin I., John M. Findlay, and James C. Bard. *A Cultural Resource Overview of the Bureau of Land Management Coleville, Bodie, Benton and Owens Valley Planning Units, California*. Oakland, Calif.: Basin Research Associates, 1979.

California Development Board. "Agricultural and Industrial Survey of Inyo County, Calif." San Francisco, September 1917. Typescript at the Eastern California Museum, Independence.

Chalfant, W. A. *The Story of Inyo*. Bishop, Calif.: Chalfant Press, 1933.

Clausen, J. C. "Report of the Owens Valley, California." November 1904. Typescript in the office of the City of Los Angeles, Metropolitan Water District.

Conkling, Harold. "Report on Owens Valley Project, California." U.S. Bureau of Reclamation, Department of the Interior, September 1921. Typescript in the office of the City of Los Angeles, Department of Water and Power.

Cotton, J. S. "Agricultural Conditions of Inyo County, California." 1905. Typescript at the Eastern California Museum, Independence.

Davidson, J. W. *The Expedition of Capt. J. W. Davidson From Fort Tejon to the Owens Valley in 1859*. Edited by Philip J. Wilke and Harry W. Lawton. Socorro, N.M.: Ballena Press, 1976.

DeDecker, Mary. "Owens Valley, Then and Now." In *Mountains to Desert: Selected Inyo Readings*, 7–15. Independence: Friends of the Eastern California Museum, 1988.

Earl, Guy Chaffee. *The Enchanted Valley and Other Sketches*. Glendale, Calif.: Arthur H. Clark Company, 1976.

EIP Associates. *Water From the Owens Valley to Supply the Second Los Angeles Aqueduct: 1970 to 1990; 1990 Onward Pursuant to a Long Term Groundwater Management Plan*. Vol. 1: *Draft Environmental Impact Report*. City of Los Angeles, Department of Water and Power; and County of Inyo, September 1990.

Faulkner, Frederick. "Owens Valley: Where the Trail of the Wrecker Runs." *Sacramento Union*, March 28–April 2, 1927.

Fox, Feramorz Young. "The Mormon Land System: A Study of the Settlement and Utilization of Land Under the Direction of the Mormon Church." Ph.D. diss., Northwestern University, 1932.

Frederick, Kenneth D. *Water for Western Agriculture*. Washington, D.C.: Resources for the Future, 1982.

Ganoe, John T. "The Origin of a National Reclamation Policy." *Mississippi Valley Historical Review* 18 (June 1931): 34–52.

———. "The Desert Land Act in Operation, 1877–1891." *Agricultural History* 10 (January 1937): 142–57.

Gates, Paul W. "The Homestead Law in an Incongruous Land System." *American Historical Review* 41 (July 1936): 652–81.

———. *History of Public Land Law Development*. With a chapter by Robert W. Swenson. Washington, D.C.: Government Printing Office, 1968.

———. "Public Land Disposal in California." *Agricultural History* 49 (January 1975): 158–78.

Hart, John Fraser. "The Middle West." *Annals of the Association of American Geographers* 62 (June 1972): 258–82.

Hill, Mary. *Geology of the Sierra Nevada*. Berkeley: University of California Press, 1975.

Hoffman, Abraham. "Joseph Barlow Lippincott and the Owens Valley Controversy: Time for Revision." *Southern California Quarterly* 54 (Fall 1972): 239–54.

———. "Origins of a Controversy: The Reclamation Service and the Owens Valley–Los Angeles Water Dispute." *Arizona and the West* 19 (Winter 1977): 333–46.

———. "Fact and Fiction in the Owens Valley Water Controversy." In *Los Angeles Westerners Corral Brand Book No. 15*, 179–91. Los Angeles: The Westerners, 1978.

———. *Vision or Villainy: Origins of the Owens Valley–Los Angeles Water Controversy*. College Station: Texas A&M Press, 1981.

Hornbeck, David. "The Far West, 1840–1920." In *North America: The Historical Geography of a Changing Continent*, edited by Robert D. Mitchell and

Paul A. Groves, 279–98. Totowa, N.J.: Rowman & Littlefield, 1987.

Hunt, Charles B. *Natural Regions of the United States and Canada.* San Francisco: W. H. Freeman and Company, 1974.

Hunter, Milton R. *Brigham Young the Colonizer.* Salt Lake City: Deseret News Press, 1940.

Ingram, Helen. "Water Rights in the Western States." In *Food Policy and Farm Programs,* edited by Don F. Hadwiger and Ross B. Talbot, 134–43. *Proceedings of the Academy of Political Science* 34 (1982).

Inyo County, California, Anno Domini 1912. Bishop, Calif.: Inyo County Register, 1912.

Inyo Independent. Independence, Calif. 1870–1924.

Inyo Register. Bishop, Calif. 1885–1925.

Jackson, Richard H. "Mormon Perception and Settlement." *Annals of the Association of American Geographers* 68 (September 1978): 317–34.

Kahrl, William L. "The Politics of California Water: Owens Valley and the Los Angeles Aqueduct, 1900–1927." *California Historical Quarterly* 55 (Spring, Summer 1976): 2–25, 98–120.

———. *Water and Power: The Conflict Over Los Angeles' Water Supply in the Owens Valley.* Berkeley: University of California Press, 1982.

Kersten, Earl W., Jr. "The Early Settlement of Aurora, Nevada, and Nearby Mining Camps." *Annals of the Association of American Geographers* 54 (December 1964): 490–507.

Kinsey, Don J. *The Water Trail: The Story of Owens Valley and the Controversy Surrounding the Efforts of a Great City to Secure the Water Required to Meet the Needs of an Ever-Growing Population.* City of Los Angeles, Department of Water and Power, 1928.

Knopf, Adolph, and Edwin Kirk. *A Geologic Reconnaissance of the Inyo Range and the Eastern Slope of the Southern Sierra Nevada, California.* U.S. Geological Survey Professional Paper 110. Washington, D.C.: Government Printing Office, 1918.

Kollmorgen, Walter M. "The Woodsman's Assaults on the Domain of the Cattleman." *Annals of the Association of American Geographers* 59 (June 1969): 215–39.

Larsen, L. T. "Hardwood Planting in Owens Valley, California." July 1, 1914. Typescript in the City of Los Angeles, Department of Water and Power, Bishop Office. Typescript.

Lawton, Harry W., Philip J. Wilke, Mary Dedecker, and William M. Mason. "Agriculture Among the Paiute of Owens Valley." *Journal of California Anthropology* 3 (Summer 1976): 13–50.

Lee, Charles H. *An Intensive Study of the Water Resources of a Part of Owens Valley, California.* U.S. Geological Survey Water-Supply Paper 294. Washington, D.C.: Government Printing Office, 1912.

Lee, Willis T. *Geology and Water Resources of Owens Valley, California.* U.S.

Geological Survey Water-Supply and Irrigation Paper No. 181. Washington, D.C.: Government Printing Office, 1906.

Lippincott, Joseph B. "The Reclamation Service in California." *Forestry and Irrigation* 10 (April 1904): 162–69.

Los Angeles. Board of Public Service Commissioners. *Owens Valley and the Los Angeles Water Supply.* 1925.

——. Department of Water and Power. "Land and Water Right Purchases: Estimate of Expenditures, Fiscal Years 1933–34 to 1936–37, Inclusive." Los Angeles, Calif. Typescript.

——. Department of Water and Power. Aqueduct Investigation Board. "Report of the Aqueduct Investigation Board to the City Council of Los Angeles." August 31, 1912. Typescript.

McCarthy, W. R., and J. C. Clausen. "Engineers' Report on the Owens Valley Irrigation District, Inyo County, California." Bishop, California, March 20, 1923. Typescript in the office of the City of Los Angeles, Department of Water and Power.

McClure, W. F. *Letter of Transmittal and Report of W. F. McClure, State Engineer, Concerning the Owens Valley–Los Angeles Controversy to Governor Friend Wm. Richardson.* Sacramento: California State Printing Office, 1925.

McGrath, Roger D. *Gunfighters, Highwaymen and Vigilantes: Violence on the Frontier.* Berkeley: University of California Press, 1984.

McIntosh, C. Barron. "Use and Abuse of the Timber Culture Act." *Annals of the Association of American Geographers* 65 (September 1975): 347–62.

——. "Patterns from Land Alienation Maps." *Annals of the Association of American Geographers* 66 (December 1976): 570–82.

Mayo, Morrow. *Los Angeles.* New York: Alfred A. Knopf, 1933.

Meinig, D. W. *On the Margins of the Good Earth: The South Australian Wheat Frontier, 1869–1884.* Chicago: Rand McNally & Company, 1962.

——. "The Mormon Culture Region: Strategies and Patterns in the Geography of the American West, 1847–1964." *Annals of the Association of American Geographers* 55 (June 1965): 191–220.

——. *The Great Columbia Plain: A Historical Geography, 1805–1910.* Seattle: University of Washington Press, 1968.

——. *The Shaping of America: A Geographical Perspective on 500 Years of History.* Vol. 1: *Atlantic America, 1492–1800,* and vol. 2: *Continental America, 1800–1867.* New Haven: Yale University Press, 1986–.

Miller, Gordon R. "Los Angeles and the Owens River Aqueduct." Ph.D. diss., Claremont Graduate School, 1978.

Moody, Charles A. "Los Angeles and the Owens River." *Out West* 23 (October 1905): 417–42.

Mulholland, C. *Inyo County: Its Lands, Water, Soil, Climate, Mines, Scenery and Other Resources.* Los Angeles: Times-Mirror Printing and Binding House, 1893.

Nadeau, Remi. *The Water Seekers*. Garden City, N.Y.: Doubleday & Company, 1950.

——. "The Water War." *American Heritage* 13 (December 1961): 30–35, 103–7.

Nash, Gerald D. "The California State Land Office, 1858–1898." *Huntington Library Quarterly* 27 (August 1964): 347–56.

Nelson, Lowry. *The Mormon Village: A Pattern and Technique of Land Settlement*. Salt Lake City: University of Utah Press, 1952.

Newell, F. H. "The Reclamation Service and the Owens Valley." *Out West* 23 (October 1905): 454–61.

Official Records of the Union and Confederate Armies in the War of the Rebellion. Vol. 50, pt. 1. Washington, D.C.: Government Printing Office, 1897.

Okada, Yasuo. *Public Lands and Pioneer Farmers*. New York: Arno Press, 1979.

"The Owens Valley Controversy, By a Special Correspondent." *Outlook*, July 13, 1927, pp. 341–43.

Owens Valley Improvement Company. "Fortunes in Apples in Owens Valley, Inyo County, California." Typescript in the Inyo Free Library, Independence, Calif.

Pakiser, L. C., M. F. Kane, and W. H. Jackson. *Structural Geology and Volcanism of Owens Valley Region, California: A Geophysical Survey*. U.S. Geological Survey Professional Paper 438. Washington, D.C.: Government Printing Office, 1964.

Patterson, Edna B., Louise A. Ulph, and Victor Goodwin. *Nevada's Northeast Frontier*. Sparks, Nev.: Western Printing and Publishing Company, 1969.

Preston, William L. *Vanishing Landscapes: Land and Life in the Tulare Lake Basin*. Berkeley: University of California Press, 1981.

Reisner, Marc. *Cadillac Desert: The American West and Its Disappearing Water*. New York: Penguin Books, 1987.

Robinson, W. W. *Land in California*. Berkeley: University of California Press, 1948.

Saliba, Bonnie Colby. "Do Water Markets 'Work'? Market Transfers and Trade-offs in the Southwestern States." *Water Resources Research* 23 (July 1987): 1113–22.

Saliba, Bonnie Colby, and David B. Bush. *Water Markets in Theory and Practice: Market Transfers, Water Values, and Public Policy*. Boulder, Colo.: Westview Press, 1987.

Sauder, Robert A. "Sod Land Versus Sagebrush: Early Land Appraisal and Pioneer Settlement in an Arid Intermountain Frontier." *Journal of Historical Geography* 15 (October 1989): 402–19.

——. "Patenting an Arid Frontier: Use and Abuse of the Public Land Laws in Owens Valley, California." *Annals of the Association of American Geographers* 79 (December 1989): 544–69.

——. "The Agricultural Colonization of a Great Basin Frontier: Economic Organization and Environmental Alteration in Owens Valley, California, 1860–1925." *Agricultural History* 64 (Fall 1990): 78–101.

———. "Powell's Vision of Arid Land Settlement Reexamined: Owens Valley, California." *Yearbook of the Association of Pacific Coast Geographers* 52 (1990): 65–90.

———. "Owens Valley's Abandoned Landscapes." *California Geographer* 32 (1992): 61–76.

———. "Mountains and Lowlands: Patterns of Human Adjustment in Owens Valley, California." In *The Mountainous West: Explorations in Historical Geography,* edited by William Wyckoff and Lary Dilsaver. Lincoln: University of Nebraska Press, forthcoming.

Sauder, Robert A., and Rose M. Sauder. "The Morrill Act's Influence on Public Land Disposal After 1870." *Agricultural History* 61 (Spring 1987): 34–49.

Schallert, Edwin. "The Valley of Flowing Waters." *West Coast Magazine* 11 (November 1911): 135–49.

Simpson, Capt. J. H. "Journal of Mr. Edward M. Kern of an Exploration of Mary's or Humboldt River, Carson River, Carson Lake, and Owens River and Lake in 1845." In *Explorations Across the Great Basin Territory of Utah for a Direct Wagon Route from Camp Floyd to Genoa, in Carson City.* Appendix Q. Washington, D.C.: Government Printing Office, 1876.

Stegner, Page. "Water and Power." *Harper's* 262 (March 1981): 61–70.

Steward, Julian. "Irrigation Without Agriculture." In *Papers of the Michigan Academy of Science, Arts and Letters,* 12:149–56. Ann Arbor: University of Michigan Press, 1930.

———. *Ethnography of the Owens Valley Paiute.* University of California Publications in American Archaeology and Ethnography, vol. 33. Berkeley: University of California Press, 1933.

———. *Basin-Plateau Aboriginal Sociopolitical Groups.* American Ethnology Bulletin 120. Washington, D.C.: Government Printing Office, 1938.

"The Story of an Irrigated Farm." *Sierra Magazine* 2 (January 1909).

"A Theft in Water." Pts. 1–8, *Inyo Magazine* (September 1–December 15, 1908); Pts. 9–10, *Sierra Magazine* (January–February 1909).

Thomas, George. *The Development of Institutions Under Irrigation, With Special Reference to Early Utah Conditions.* New York: Macmillan Company, 1920.

Udall, James R. "Just Add Water Marketing." *Sierra* 72 (March/April 1987): 37–42.

U.S. Bureau of the Census. Ninth Census, 1870. Manuscript Census Schedules of Inyo County, California. Washington, D.C.: National Archives. Microfilm.

———. Tenth Census, 1880. Manuscript Census Schedules of Inyo County, California. Washington, D.C.: National Archives. Microfilm.

———. *Tenth Census, 1880: Report Upon the Statistics of Agriculture* Washington, D.C.: Government Printing Office, 1883.

———. *Report of the Statistics of Agriculture in the United States at the Eleventh Census, 1890.* Washington, D.C.: Government Printing Office, 1895.

——. *Twelfth Census of the United States Taken in the Year 1900. Agriculture,* pts. 1 and 2. Washington, D.C.: Government Printing Office, 1902.

——. *Thirteenth Census of the United States, Taken in the Year 1910.* Vol. 6: *Agriculture 1909 and 1910.* Washington, D.C.: Government Printing Office, 1913.

——. *Fourteenth Census of the United States, Taken in the Year 1920.* Vol. 6, pt. 3: *Agriculture: The Western States and Outlying Possessions.* Washington, D.C.: Government Printing Office, 1922.

——. *United States Census of Agriculture, 1925.* Pt. 3: *The Western States.* Washington, D.C.: Government Printing Office, 1928.

U.S. Congress. Senate. *Survey of Conditions of the Indians in the United States.* Pt. 29: *California.* Hearings Before a Subcommittee on Indian Affairs, United States Senate. 72d Congress, 1st Session, September 20–28, 1932. Washington, D.C.: Government Printing Office, 1934.

——. *Report of the Special Committee of the United States Senate on the Irrigation and Reclamation of Arid Lands.* 2 Vols. 51st Cong., 1st Sess. Senate Report 928, pt. 3, vol. 2: *The Great Basin Region and California.* Washington, D.C.: Government Printing Office, 1890.

U.S. Department of the Interior. Bureau of Land Management. Master title plats. Sacramento. Microfilm.

——. Historical indexes. Sacramento. Microfilm.

——. Independence District Land Office tract books. Sacramento. Microfilm.

U.S. General Land Office. *Instructions to the Surveyors General of Public Lands of the United States for Those Surveying Districts Established in and Since the Year 1850; Containing Also a Manual of Instructions to Regulate the Field Operations of Deputy Surveyors.* Washington, D.C.: Government Printing Office, 1871.

U.S. Geological Survey. *Second Annual Report of the Reclamation Service, 1902–3.* 58th Cong., 2d Sess., House Doc. 44. Washington, D.C.: Government Printing Office, 1904.

——. Bishop Quadrangle. 1:125,000. 1913.

——. Mt. Whitney Quadrangle. 1:125,000. 1907.

Von Schmidt, A. W. Field Notes. Sacramento: U.S. Department of Interior, Bureau of Land Management. Microfilm.

Wagner, W. F., ed. *Leonard's Narrative: Adventures of Zenas Leonard, Fur Trader and Trapper, 1831–1836.* Cleveland: Burrows Brothers Company, 1904.

Wahl, Richard W., and Frank H. Osterhoudt. "Voluntary Transfers of Water in the West." In *National Water Summary: 1985,* 113–24. U.S. Geological Survey Water-Supply Paper 2300. Washington, D.C.: Government Printing Office, 1986.

Walter, Nancy Peterson. "The Land Exchange Act of 1937: Creation of the Indian Reservations at Bishop, Big Pine, and Lone Pine, California, Through a Land Trade Between the United States of America and the City of Los Angeles." Ph.D. diss., Union Graduate School, 1986.

Walton, John. *Western Times and Water Wars: State, Culture, and Rebellion in California.* Berkeley: University of California Press, 1992.

———. "Picnic at Alabama Gates: The Owens Valley Rebellion, 1904–1927." *California History* 65 (September 1986): 192–207.

Watson, E. B., and R. Earl Storie. *Soil Survey of the Bishop Area, California.* U.S. Department of Agriculture Series 1924, no. 3. Washington, D.C.: Government Printing Office, 1928.

Weatherford, Gary D., and Steven J. Shupe. "Reallocating Water in the West." *American Water Works Association Journal* 78 (October 1986): 63–71.

Wentworth, Edward Norris. *America's Sheep Trails: History, Personalities.* Ames: Iowa State College Press, 1948.

Whitney, J. D. "The Owen's Valley Earthquake. In Two Papers. 1. Local Details." *Overland Monthly* 9 (August 1872): 130–40.

Wilkinson, Charles F. "Western Water Law in Transition." *American Water Works Association Journal* 78 (October 1986): 34–47.

Wood, Richard Coke. *The Owens Valley and the Los Angeles Water Controversy: Owens Valley As I Knew It.* Stockton, Calif.: Pacific Center for Western Historical Studies, University of the Pacific, 1973.

Worster, Donald. *Rivers of Empire: Water, Aridity and the Growth of the American West.* New York: Pantheon Books, 1985.

Yonay, Ehud. "How Green Was My Valley." *New West* 2 (March 28, 1977): 21–25.

Index

About the Author

Robert A. Sauder is Professor of Geography at the University of New Orleans. Born and raised in Southern California, he has long held an interest in the Owens Valley–Los Angeles water conflict. After receiving his undergraduate degree in geography from California State University at Long Beach, he attended the University of Oregon, completing his doctorate in 1973. He has published numerous articles in national and international journals on the historical geography of the West. His preoccupation with the arid West serves as an antidote to more than two decades of life in the steamy swamplands of southern Louisiana.